T0331173

Generative AI and Digital Forensics

In today's world, cybersecurity attacks and security breaches are becoming the norm. Unfortunately, we are not immune to it, and any individual or entity is at dire risk. The best and only thing that we can do is to mitigate the risks as much as possible so that they do not happen at all. But even when a security breach does indeed happen, the immediate reaction is to contain it so that it does not penetrate further into the information technology/network infrastructure. From there, mission-critical processes need to be restored, until the business can resume a normal state of operations, like it was before the security breach.

But another key step here is to investigate how and why the security breach happened in the first place. The best way to do this is through what is known as "digital forensics". This is where specially trained digital forensics investigators collect and comb through every piece of evidence to determine this. Eventually, the goal is then to use this evidence in a court of law so the cyberattacker can be made to answer for their crime and eventually be brought to justice.

However, the area of digital forensics is a large one, and many topics around it can be covered. Also, generative AI is being used to not only help in the analysis of the evidence but also to help automate the digital forensics software packages that are available today. Therefore, in this book, the following is covered:

- Examples of security breaches and overview of digital forensics.
- How digital forensics can be used to investigate the loss or theft of data.
- An introduction to the SQL Server Database.
- A review of the SQL Injection Attack.
- How generative AI can be used in digital forensics.

Ravindra Das is a technical writer in the cybersecurity realm. He also does cybersecurity consulting on the side through his private practice, ML Tech, Inc. He holds the Certified in Cybersecurity certification from the ISC(2).

Generative AI and Digital Forensics

Ravindra Das

CRC Press
Taylor & Francis Group
Boca Raton London New York

CRC Press is an imprint of the
Taylor & Francis Group, an **informa** business

Designed cover image: © Shutterstock

First edition published 2025
by CRC Press
2385 NW Executive Center Drive, Suite 320, Boca Raton FL 33431

and by CRC Press
4 Park Square, Milton Park, Abingdon, Oxon, OX14 4RN

CRC Press is an imprint of Taylor & Francis Group, LLC

© 2025 Ravindra Das

Library of Congress Cataloging-in-Publication Data
Names: Das, Ravindra, author.
Title: Generative AI and digital forensics / Ravindra Das.
Other titles: Generative artificial intelligence
Description: First edition. | Boca Raton : CRC Press, 2025. |
Includes bibliographical references and index.
Identifiers: LCCN 2024029334 (print) | LCCN 2024029335 (ebook) |
ISBN 9781032742458 (hardback) | ISBN 9781032744865 (paperback) |
ISBN 9781003469438 (ebook)
Subjects: LCSH: Computer security. | Digital forensic science. |
Artificial intelligence. Classification: LCC QA76.9.A25 D345 2025 (print) |
LCC QA76.9.A25 (ebook) | DDC 005.8--dc23/eng/20240821
LC record available at https://lccn.loc.gov/2024029334
LC ebook record available at https://lccn.loc.gov/2024029335

ISBN: 978-1-032-74245-8 (hbk)
ISBN: 978-1-032-74486-5 (pbk)
ISBN: 978-1-003-46943-8 (ebk)

DOI: 10.1201/9781003469438

Typeset in Adobe Caslon Pro
by KnowledgeWorks Global Ltd.

This book is dedicated to my Lord and Savior, Jesus Christ,
the Grand Designer of the Universe,
my parents, Dr. Gopal Das and Mrs. Kunda Das,
and my loving cats, Fifi and Bubu.
This book is also dedicated to: Richard and Gwynda Bowman,
Jaya Chandra, Tim Auckley, Patricia Bornhofen, and Ashish Das.

Contents

Acknowledgments

I would like to thank Ms. Gabriella Williams, my editor, who made this book a reality.

SECURITY BREACHES AND OVERVIEW OF DIGITAL FORENSICS

There is no doubt today that the world of cybersecurity is becoming a very broad and complex one. Gone are the days when it was a term that nobody could even conjure up. Even when the first phishing attack hit America Online (AOL) back in the late 1990s, during the height of the ".com" bubble, nobody ever really gave a second thought to it. It was expected that it would happen just that one time, but fast forward some 30 years later, now it is all across the news headlines.

Another typical example of this is the critical infrastructure of the United States. This typically includes the nation's electric grid, water supply, food distribution system, nuclear facilities, rail lines, etc. When these pieces of infrastructure were designed back in the 1960s and 1970s, the main concern at that time was purely about physical access entry. The issues of cybersecurity at that time were not even conceived of. In fact, the next section outlines some of the top attacks that have occurred to critical infrastructure, on a global basis.

The Top Attacks

1. Attacks on the power grids in the Ukraine:
   This occurred in December 2015. The electric grid still made use of the traditional supervisory control and data acquisition (SCADA) system, which was not upgraded for the longest time. This cyberattack impacted about 230,000 residents in that area who were without power for a few hours. Although this threat variant was short-lived, it further illustrates the grave weaknesses of the critical infrastructure. For example, the traditional spear phishing email was used to launch the threat vector, and in fact, just a year later, the same email

DOI: 10.1201/9781003469438-1

was used to attack an electrical substation near Kyiv, causing major blackouts for a long period of time.

2. Attack on the water supply lines in New York:
   The target this time was the Rye Brook Water Dam. Although the actual infrastructure was small in comparison, the lasting repercussions were magnanimous. The primary reason for this is that this was one of the first instances in which a nation-state actor was actually blamed, and all fingers pointed toward Iran. The most surprising facet of this cyberattack was that it was not reported to law enforcement until 2013.agencies until 2013. Even more striking is that the malicious threat actors were able to gain access to the command center of these facilities by using just an ordinary dial-up modem.

3. Impacts on the ACH system:
   Although the global financial system may not directly fit into the classical definition of a critical infrastructure, the impacts felt by any cyberattack can be just as great. In this threat variant, it was the SWIFT Global Messaging system that was the primary target. This is used by banks and other money institutions to provide details about the electronic movement of money which includes ACH, wire transfers, etc. This is a heavily used system worldwide, as almost 34 million electronic transfers make use of this particular infrastructure. The Lazarus Cyberattack group, originating from North Korea, was able to gain a foothold in the banks by using hijacked SWIFT login username and password combinations. This attack has been deemed to be one of the first of its kind in the international banking sector.

4. Damages to nuclear facilities:
   Probably one of the well-known cyberattacks on this kind of infrastructure was upon the Wolf Creek Nuclear Operating Corporation, which is located in Kansas. In this instance, spear phishing emails were leveraged against the key personnel working at this facility, who had specific control and access to the controls. Although the extent of the damage has been kept classified, this situation demonstrates clearly just how vulnerable the US-based nuclear facilities are. For example, if a cyberattacker were to gain access to one, they

could move in a lateral fashion to other nuclear power plants, causing damage in a cascading style, with the same or even greater effects of that caused by a Thermonuclear War.

5. Attack on the water supply:

The most well-known attack just happened recently in Oldsmar, Florida. Although the details of this cyberattack are still coming to light, it has been suspected that the hacker was able to gain control by making use of a remote access tool, such as TeamViewer. But there were other grave weaknesses in the infrastructure as well, such as a very outdated operating system (OS) and poor password enforcement (such as not creating long and complex ones and rotating them out on a frequent basis). In this instance, the goal of the cyberattacker was not just to cause damage to the water supply system but to even gravely affect the health of the residents who drank the water, by poisoning it with a chemical-based lye. Luckily, an employee was able to quickly notice what was going on and immediately reversed the settings that were out into motion by the cyberattacker. However, it is still not known whether this hack occurred outside the US soil or from within. If it is the latter, then this will raise even more alarm bells that domestic-based cyberattackers are just as much of a grave threat as the nation-state actors to our critical infrastructure.

It is important to note that many efforts have been made to try to make the nations' critical infrastructure as cyber-proof and resilient as possible. But the main problem is that since the technologies that were used to create them are so outdated, it is almost impossible to find any replacement parts for them. In fact, many of the vendors that provided these key pieces of technology are now no longer in business. If parts do have to be procured, they will most likely have to be sought from a different country, namely, China.

Another key issue with trying to upgrade a critical infrastructure is that one simply cannot upgrade with new software patches and firmware because it has to be interoperable with the older technologies that are present. And given the extremes of the two, it would be very difficult to first test them in a "sandbox"-like environment before they are released into the production environment. In this regard, two of the oldest technologies that have been deployed are known as the "SCADA" and the "ICS". These are further reviewed in the next two sections.

The Security Issues of SCADA

SCADA is an acronym that stands for "supervisory control and data acquisition". The technical definition of it is as follows:

> (SCADA) systems are used for controlling, monitoring, and analyzing industrial devices and processes. The system consists of both software and hardware components and enables remote and on-site gathering of data from the industrial equipment. In that way, it allows companies to remotely manage industrial sites such as wind farms, because the company can access the turbine data and control them without being on site.
>
> *https://scada-international.com/what-is-scada/*

There are key security issues with SCADA, and the major ones are as follows:

- Outdated technologies:
  Many of the SCADA systems that are in use today have been deployed decades ago. The major concern now is that the SCADA system will be used as a point of entry to launch an attack on a critical infrastructure.
- Open visibility:
  Many systems are in open and because of that, there are greater chances of an insider attack. There is a growing awareness in this aspect, and businesses that make use of SCADA are trying to put advanced physical controls in place to protect it. But the main problem is that these newer technologies have to be layered onto the existing legacy security system which is in place.
- Network integration:
  SCADA systems were designed to operate by themselves, meaning any future integration into other technologies was not even considered. With the advent of the Internet of Things

(IoT), everything is now interconnected with each other, even the SCADA systems. Once again, there are interoperability issues that are coming out, and this increased interlinking is also expanding the attack surface for the cyberattacker.

How to Address the Security Issues of a SCADA System

As mentioned, the main issue with SCADA systems is that a bulk of them were built in the 1970s and 1980s. Because of this, you simply cannot "rip out" the old and put in newer technology in order to secure it.

But, in the end, it is still possible to secure SCADA systems, and here are some ways in which it can be done:

1. Correctly ascertain all of the connections to the SCADA system. This is like conducting a risk assessment for information technology (IT)/network infrastructure.
2. If there are any connections that are deemed to be not needed, disconnect them all at once. This is like network ports when they are not being used.
3. For the connections that are absolutely needed, make sure that they are hardened to the greatest extent possible.
4. Although SCADA systems have been built with proprietary technologies that are not designed to co-mingle with others, do not further implement any proprietary protocols.
5. Determine to see if there are any hidden backdoors in the system.
6. It is important to deploy firewalls, network intrusion devices, and routers surrounding the SCADA system so that you can be notified in real time of any potential security breaches that may be happening. Also, make use of a $24 \times 7 \times 365$ incident monitoring tool.
7. On a regular basis, conduct risk assessments and audits of all internal and remote devices that are connected to the SCADA system.
8. Again, just as you would for your IT/network infrastructure, it is important to define to roles and responsibilities as to whom will actually "protect" the SCADA system.

The Cyber Risks That Can Potentially Affect an ICS

ICS is an acronym that stands for "industrial control systems". The technical definition of it is as follows:

> These systems consist of equipment and software that monitor and control physical processes in various sectors, such as manufacturing, industrial production, energy management, water treatment, transportation and healthcare. ICS is essential for the efficient and safe operation of many critical infrastructures and services that we rely on every day.
>
> *https://www.mwes.com/types-of-industrial-control-systems/*

Here are some of the securities that are associated with an ICS:

1. Air gapping will no longer work:
   One of the most extensive security measures that could be afforded during those times is what is known as "air gapping". In a way, this is very similar to dividing up your IT/network infrastructure into different regimes, also known as "subnets". With air gapping, the ICS network was wholly isolated from the rest of the critical infrastructure.

   The theory was that if an insider attack were launched, any effects from it would not be transmitted down to the ICS system. But even now, air gapping is not a feasible solution to protect against cyberattacks. The primary reason for this is that both the physical and digital/virtual worlds are now coming together and joining as one whole unit through a phenomenon called the "industrial internet of things" (IIoT).

2. Legacy hardware and software components:
   Because of the major difficulties in finding the right security tools to add on, many critical infrastructures are still using outdated hardware and software components. Among the most at risk to a cyberattack are the following:
   - Programmable logic controllers (PLCs)
   - Remote terminal units (RTUs)
   - Distributed control systems (DCSs)

   The devices mentioned above are typically used to manage the processes and the subprocesses of the ICS network.

Since the flow of network communications is done in a clear text format from within the ICS network, a cyberattacker could be literally on the other side of the world and deliver their malicious payload to an oil refinery in the southern United States. But worst yet, many of the OSs used in critical infrastructure are outdated and even no longer supported by Microsoft.

3. There is no clear-cut visibility:

An ICS offers no physical visibility, and as a result, it is almost too hard to detect if there is any suspicious behavior that is transpiring until it is way too late. Thus, many of the settings in an ICS are difficult to configure properly to meet today's demand for the basic utility necessities of the everyday American.

4. The communications protocols are outdated:

Most of the ICS networks of today are deemed to be outdated in nature. For example, this is most prevalent in the so-called "control layer" protocols that are used. For example, the mathematical logic implemented into the ICS hardware can be easily changed around, resulting in an unintentional flow in mission-critical operations.

What the Future Holds for Critical Infrastructure

Now that we have reviewed what some of the major threats are to critical infrastructure, as well as its weaknesses from the standpoint of the SCADA and ICS systems, what does the future of it look like? Here are some thoughts on this topic:

1. Segmentation could occur:

With the implementation of multi-factor authentication (MFA) and the zero trust framework, there have been calls now to further divide up the IT and network infrastructure that exists in the internal environment into smaller chunks, and this is known as "segmentation". Each segment would have its own set of defenses, and the statistical probability of a cyberattacker breaking through all of these segments becomes lower each and every time, and as a result, they give up in frustration.

2. The IoTs:

There is a great interest and even efforts are currently being undertaken to bring the world of the IoT into critical infrastructure. This is known as the "industrial internet of things" (IIoT). The reason for this is simple: with an IIoT in place, the attack surface becomes much greater, and the number of backdoors that the cyberattacker can penetrate into is now greatly multiplied.

3. The financial damage will escalate:

The financial toll that it will take on critical infrastructure that is impacted is expected to reach well over the multimillion-dollar mark. Also, with the convergence that is currently taking place within the IT and the operational technology (OT) technologies, the cyberattacker will be able to easily gain access to either the ICS or SCADA systems via any vulnerabilities or gaps that still persist in the network of the critical infrastructure.

4. A closer collaboration with cybersecurity:

There will be a new movement that has been termed appropriately the era of "shared responsibility". This is where the leaders of critical infrastructure will start to work closely with the cybersecurity industry.

5. A greater need for cybersecurity insurance:

Leaders of critical infrastructure are starting to understand the need for some sort of financial protection in case they are breached. Thus, there will be a great increase in demand for cybersecurity insurance policies in the coming years.

6. A migration to the cloud:

A 100% migration will probably not happen for critical infrastructure. The primary reason for this is that once again, most of the technologies that were developed for critical infrastructure were developed back in the 1970s and 1980s.

Vulnerability and Exploitation

The reason why we have picked critical infrastructure as the prime example of weaknesses is that it is important to showcase that nothing is immune to a cyberattack, or even a security breach.

In other words, all assets that a company may possess, whether it is digital or physical in nature, are prone to being vulnerable. As we have also described in our previous publications, one of the best ways to find just how vulnerable they are is to conduct a risk assessment analysis.

With this methodology in hand, you and your IT security team will work together to take an inventory of all of the assets (digital and physical) that exist within your business. Once you have done this, then the next step is to break down in which categories they fall under. Once this has been accomplished, you then need to take each category and all of the assets in them and determine just how vulnerable they truly are.

By this, we mean what kinds of gaps and weaknesses are prevalent in them, and just how likely they will be exploited if a security breach were to actually happen. In this regard, the terms "vulnerability" and "exploitation" are often used together. The truth of the matter is that the two are very different, and thus the technical definitions for them are as follows:

Vulnerability can be defined as:

A vulnerability, in information technology (IT), is a flaw in code or design that creates a potential point of security compromise for an endpoint or network. Vulnerabilities create possible attack vectors, through which an intruder could run code or access a target system's memory.

https://www.techtarget.com/whatis/definition/vulnerability

Exploitation can be defined as follows:

The use of actions and operations—perhaps over an extended period of time—to obtain information that would otherwise be kept confidential and is resident on or transiting through an adversary's computer systems or networks.

https://www.infosecinstitute.com/resources/general-security/
cyber-exploitation/

To put in simpler terms, a "vulnerability" is simply the opening that is made available for a cyberattacker to penetrate into. "Exploitations" is when the cyberattacker actually takes advantage of that hole or gap and uses that in order to deploy a malicious payload, with the main intention to cause some sort of damage to the business entity, such as a data exfiltration attack.

A good example of this is the database. This is a software application that is used quite a bit in order to house the information and data that a business makes use of on a daily basis. There are many different kinds and types of databases that are available, such as those of Oracle, SQL Server, MySQL, PostgreSQL, etc. But when the source code is created for a database application, it is often not checked for any kinds or types of vulnerabilities. Thus, the cyberattacker can exploit this and insert the malicious code.

As a precursor, this chapter and the next will actually involve the study of a database exploitation attack using a SQL Database. So in this example, if there is a gap in this particular SQL Database, the cyberattacker can easily penetrate this and launch what is known as a "SQL Injection Attack". This is yet another concept that will also be explored in this book. This is where malicious code is actually deployed in order to alter the contents that reside in the SQL Server Database.

An example of a SQL is illustrated in Figure 1.1.

Figure 1.1   An example of a SQL database.

(From https://www.shutterstock.com/image-illustration/relational-database-tables-on-databases-placed-2258134001)

Table 1.1 A risk rating scale.

RANKING	DEGREE OF SEVERITY
1	Least vulnerable
2–4	More vulnerable
5	Moderately vulnerable
6–9	Increasing vulnerability
10	Most, or severe vulnerability

Once you have discovered the vulnerability of both the digital assets and the physical across all of the categories, then the next step is to assign them a numerical ranking in order to demonstrate their degree of severity of vulnerability. For example, and from the standpoint of simplicity, you can assign a numerical scale from 1 to 10, where "1" would be the "least vulnerable" and "10" would be "most vulnerable". This is further exemplified in Table 1.1.

The rule of thumb here is to either implement new controls or upgrade the existing ones that are deemed to be "most vulnerable" first, then go down in a cascading fashion to address the other digital assets and physical assets that are not as "vulnerable". But, in order to determine which of the digital assets and physical assets have gaps or weaknesses in them that make them indeed "vulnerable", you and your IT security team need to conduct a series of tests. Some of the most comprehensive ones you can use are of penetration testing and threat hunting, which are reviewed in the next section.

The Tools to Discover the Vulnerabilities

Penetration Testing

A penetration test (also known as a "pen test") can be technically defined as follows:

> Penetration testing, aka pen testing or ethical hacking, attempts to breach a system's security for the purpose of vulnerability identification.
>
> *https://www.hackerone.com/knowledge-center/*
> *what-penetration-testing-how-does-it-work-step-step*

In other words, you are literally taking down the digital assets and the physical assets of an IT/network infrastructure in order to fully determine where the weaknesses lie. This is also referred to as "ethical hacking". It is important to note that pen testing is simply not just confined to the realm of digital and physical assets. It could also include launching social engineering attacks in order to determine how easily the employees of a business can fall prey into giving up their confidential and private information.

The Types of Penetration Tests

From within the realms of pen testing, there are many specific tests that can be conducted, which include:

- Backdoors in the OS
- Unintentional flaws in the design of the software code
- Improper software configuration management implementation
- Using the actual software application in a way it was not intended to be used
- Servers
- Network endpoints
- Wireless networks
- Network security devices (this is hit upon the most in an actual pen test, which includes the routers, firewalls, network intrusion devices, etc.)
- Mobile and wireless devices
- Other areas of exposure, such as software applications and the code behind it
- Black-box testing
- White-box testing
- Gray-box testing
- Firewall configuration testing
- Stateful analysis testing
- Firewall bypass testing
- IPS evasion
- DNS attacks which include:
  - Zone transfer testing
  - Any types or kinds of switching or routing issues
  - Any other required network testing

- Eavesdropping
- Data modification
- Identity spoofing (also known as IP address spoofing)
- Password-based attacks
- Distributed denial of service attacks (DDoS)
- Man in the middle attacks
- Compromised key attacks
- Sniffer attacks
- Application layer attacks
- Web application testing
- Database testing
- Dark web testing

However, in the end, whatever needs to be tested will be entirely up to the client that is requesting these particular services. In this regard, some of the factors to take into account include the following:

1. The size of the business or corporation in question (this can be a direct function of employee size).
2. The complexity of the IT infrastructure that is to be penetration tested.
3. Any other security breaches that may have occurred to the client.
4. Any other known gaps and vulnerabilities that are known but have not been remediated yet.
5. Any other types or kinds of social engineering attacks that may have also occurred in the past.

With regard to the size of the business, for example, if the organization that wishes to have penetration testing conducted on its premises is a small one, with less than 20 employees, one can assume that the IT infrastructure is relatively simple.

Thus, in this particular instance, a complete penetration testing team may not be needed. Rather, just two or three pen testers may be needed in order to execute and complete the required tests and compile the report(s) which will summarize both the findings and recommendations.

The Penetration Testing Teams

With regard to the pen testing teams, there are three that are most commonly used (once again, this is purely dependent upon the size of the business, and the scope of the test that is requested by the client. They are as follows:

1. The Red Team:

   This team has the ultimate responsibility of launching a realistic but mock cyberattack against the defense perimeters of the client. This is in efforts in order to uncover its true security vulnerabilities, weaknesses, and holes. But apart from this, they are also interested in the access methods to get those targets. In other words, they will take the full mindset of an actual cyberattacker. Also, this kind of team will use an enormous amount of creativity and use techniques one may have never heard of. The final goal here is to breach through the lines of defenses of the client and then through each and every means that are available at their disposal. But to accomplish this, they will think and act just like a real cyber attacker, but very often come up with ideas on their own. Also, the Red Team is interested in those systems in your IT infrastructure that are "out of scope". As a result, this gives the Red Team a much broader set of parameters to examine. At this point, it is very important to note that the Red Team simply cannot pick targets. Rather, they can recommend a list of targets to the client, but in the end, the client has to give final approval. Also, the Red Team has to make the risks known to the client, and they have to fully agree to this. The Red Team will take every precaution that nothing out of the ordinary actually happens, but there is a chance that it could happen, and the client has to completely acknowledge this fact, and be prepared to deal with it in case something like this does actually happen in the end.

2. The Blue Team:

   This team can be considered literally as the "good guys", in the sense that they will work directly with CISO and the IT security team, in an effort to thwart the efforts of the

Red Team. In this regard, some of the key activities of the Blue Team include the following:

- Preparedness:

  The Blue Team will do everything in its power to protect the digital and the physical assets of the IT/network infrastructure of the client and to keep all key stakeholders apprised of what is actually happening.

- Identification:

  The members of the Blue Team will make every effort that is possible to correctly identify any potential cyberattacks that the Red Team is actually planning. It is important to note here that the two teams (the Red Team and the Blue Team) will not communicate with each other during this phase of the penetration testing exercises.

- Containment:

  If the Red Team is successful in penetrating through the lines of defenses of the client, then it will then become the primary responsibility of the Blue Team to contain the damage that is incurred. In this regard, one of the best tools that the Blue Team will have at hand is the incident response plan.

- Recovery:

  If the client has actually been breached by a cyberattack from the Red Team, it will also be the primary responsibility of the Blue Team to launch the disaster recovery and business continuity plans in order to bring the client back to a predefined level of both process and system functionality.

- Lessons learned:

  After the Red Team has launched their cyberattacks, the Blue Team will then compile all of the data that they have collected and include it in the final report for the client. This section will be entitled: "The Lessons That Have Been Learned".

  The Blue Team will also work with the CISO and the IT security team of the client in order to further harden up the OSs and the perimeter lines of defenses of the client.

3. The Purple Team:

This team can be considered as the summation of the Blue and Red Teams and serves as a "check and balance" to both teams. Some of their specific functionalities include the following:

- Working with both the Red and Blue Teams in a seamless fashion:
  This includes making observations and notes as to how the two teams are working together and making any needed adjustments to the actual penetration test.
- Understanding and visualizing the big picture:
  The Purple Team has to take the responsibilities of both the Red and Blue Teams.
- Assuming an overall responsibility for the penetration testing exercise(s):
  This simply refers to the preparation of the final report for the client, as well as analyzing and interpreting the results for them.
- Delivering the maximum value to the client:
  By collecting information and data from both the Red Team and the Blue Team, the Purple Team can create a high-caliber and informative final report for the client.

The Tools of Penetration Testing

There are many tools that a penetration testing team can use, and the following is a sampling of them:

1. The port scanner:

These kinds of tools typically gather information and data about a specific target in the network environment of the IT infrastructure, and they usually check for any network ports that are open, but either they are not known, or simply that the IT security team has forgotten to shut them down. Some examples of what can be scanned include:

- The SYN-SYN-ACK-ACK sequence for TCP ports
- Various half-scans (this is when a cyberattacker attempts to connect to a remote computer but does not send any ACK data packets in response to the SYN/ACK data packets)
- Detecting the OS type

2. The vulnerability scanner:

This kind of device attempts to find any *known* vulnerabilities in the targeted system. There are two kinds of vulnerability scanners:

- Network based:
  These only scan for the targeted OS and the network infrastructure in which they reside, as well as other TCP/IP-based devices, which may exist in this kind of environment.
- Host based:
  These can scan an entire OS for any *known* vulnerabilities and weaknesses, as well as for any software configuration problems (this includes file access/user permission management protocols). They typically scan for any gaps, weaknesses, vulnerabilities, and backdoors that may reside in the source code.

3. The application scanner:

This kind of tool examines for any security weaknesses in web-based applications, such as:

- Memory buffer overruns
- Cookie manipulations
- Malicious SQL injections
- Cross-site scripting (also known as "XSS")

4. The web application assessment proxy:

This is a tool that is deployed between the web browser and the web server, upon which the specific web application resides. As a result, all of the information and data flow between the two can be examined at a close level.

In terms of commercial off-the-shelf (also known as "COTS"), some of the most popular tools include:

- The network mapper:
  As the name implies, this is used primarily for detecting weaknesses or holes in the network environment.
  - What hosts are available.
  - The information by the hosts.
  - What OS is being used (this is also known in technical terms as "fingerprinting").
  - The versions and the types of data packet filters/firewalls are being used by any particular host.

By using NMAP, you can create a virtual map of the network segment or even the entire network infrastructure.

- Metasploit:
  This is a package of different pen testing tools and a framework that is constantly evolving in order to keep up with today's cyber threat landscape. This tool also comes with what is known as a "meterpreter", which displays the results after an exploit has occurred. As a result, this can be quickly analyzed and interpreted by the pen tester to the client, and from there, formulate the appropriate strategies for remediation and control deployment.
- Wireshark:
  Unlike NMAP, this tool is an actual network protocol and data packet analyzer which can analyze the security weaknesses of the traffic in real time. For example, live information and data can be collected from:
  - IEEE 802.11
  - Bluetooth
  - Token ring
  - Frame relay
  - IPsec
  - Kerberos
  - SNMPv3
  - SSL/TLS
  - WEP
  - Any ethernet-based connections
- John the Ripper:
  One of the biggest cybersecurity threats has been and will continue to be the inherent weaknesses of the traditional password. One of the best-known tools that can discover the weaknesses of passwords is known as "John the Ripper" (JTR). There is nothing too complex about this tool, its elegance is its simplicity in of itself. It can be used by pen testers primarily to initiate dictionary attacks. Further, a key advantage of this tool is that it can be modified to test for all the varieties of dictionary attacks that could occur.

Threat Hunting

Another tool that can be used to discover the gaps and weaknesses in both the digital assets and physical assets is what is known as "threat hunting". It can be technically defined as follows:

> Cyber threat hunting is proactively and systematically searching for signs of potential cyber threats within an organization's network or systems. This can be done through manual and automated techniques, such as analyzing log data, conducting network scans, and using threat intelligence feeds. Cyber threat hunting aims to identify potential threats that may have evaded traditional security controls, such as firewalls or intrusion detection systems.

> *https://www.sentinelone.com/cybersecurity-101/threat-hunting/*

Whereas penetration testing looks for holes in the perimeter defense by threat variants from the external environment, threat hunting looks for any signs of danger that are residing from within the confines of the IT/network infrastructure of any business.

Before any threat hunting activities can start, you will need to first conduct a risk assessment, in a manner similar to that has been described earlier in this chapter.

The Risk Assessment

There are nine distinct steps here, which are as follows:

1. Threat and vulnerability management:
   In this first phase, you are implementing a classification scheme in order to determine what is most at risk in your existing IT infrastructure. In the end, all assets need to be examined, but in this specific scenario, you are using a top-down approach, as also reviewed earlier in this chapter.
2. Identity management:
   In this part, you need to identify the employees of your business who are trying to gain access to shared network resources. For example, by making use of password managers, two-factor authentication (2FA), MFA, or technologies such as biometrics.

3. Security awareness training and education:

   With this, you need to take a careful look as to how all of the employees in your business or corporation are being trained in protecting those IT resources that they use for their daily functions. Also, you need to determine the overall level of "cyber hygiene" that they demonstrate.

4. The security policy:

   At this point, you and your IT security staff need to carefully scrutinize the existing security policy that has been crafted and implemented. You will need to determine what is outdated, what needs to be updated, or what needs to be replaced in its entirety.

5. The incident response plan:

   At this phase, the incident response plan needs to be carefully audited as well. Some things that need to be examined are how often it has been updated, and if it has been practiced in real time. Also, you will need to determine if all of the contact information of the members of the incident response team is accurate and updated with their most recent information.

6. The existing IT infrastructure:

   At this point, you and your IT security team are trying to probe for any vulnerabilities and weaknesses that currently exist in the current IT/network infrastructure.

7. Security management:

   In this part, you are carefully scrutinizing your current IT security staff, by taking a close look at who is responsible for what in the overall perimeter defense or even in the zero trust framework.

8. Emerging security technologies:

   At this phase, you are taking a careful look as to what newer security tools could be potentially implemented and/or upgraded in order to further beef up the layers of the perimeter defenses at your organization.

9. The use of third-party vendors:

   In this last phase, you need to carefully audit the selection process of hiring third-party vendors, and if efforts are currently being made to scrutinize their existing security processes to make sure that they are aligned with yours.

Threat Hunting Models

It is important to note at this point that you and your IT security team do not have to create your threat hunting when and as it is needed. There are numerous frameworks that you can use that are available from both NIST and the private sector. With respect to the latter, there is one that has been around for a long time, and it is called the "cyber kill chain model", which is developed and produced by Lockheed Martin. It consists of seven distinct phases, which are as follows:

1. Reconnaissance:
   The cyberattacker takes the needed time to research their potential target, determines their weaknesses and vulnerabilities, and calculates the best way for covert penetration.
2. Weaponization:
   This is where the cyberattacker creates their malicious payload which is to be deployed at the target, once they are in.
3. Delivery:
   The malicious payload is now launched toward the target.
4. Exploitation:
   The cyberattacker, through the malicious payload that has been deployed, now takes full advantage of the weaknesses that were found in the target.
5. Installation:
   The malicious payload creates more backdoors into the target so that the cyberattacker can move in a lateral fashion throughout the entire IT/network infrastructure.
6. Command and control:
   At this stage, the cyberattacker can now manipulate the target in a way or fashion that will cause the maximum amount of damage.
7. Actions on objective:
   Finally, the cyberattacker takes the required actions in order to reach their final, desired objectives. A prime example of this is data exfiltration.

This model is illustrated in Figure 1.2:

Figure 1.2

Another useful threat hunting framework that you can use is called "SOAR". It is an acronym that stands for the following:

- **S**ecurity **O**rchestration
- **A**utomation
- **R**esponse

With security orchestration, you and your IT security staff are bringing together all of the available threat hunting tools that you currently have, and making them all work together as one, cohesive unit. Some of the features of security orchestration include the following:

- Having a standard set of threat hunting processes.
- Providing a single platform in which the threat hunting team can input information and data as they are collected in real time.
- Providing a unified dashboard from which all alerts and warnings can be further examined, such as using a SIEM.

Some of the key advantages of using the SOAR methodology include the following:

- Examine for any potential threat variants or abnormal behavior on a 24 × 7 × 365 basis.
- Providing a centralized platform in which to further probe into hidden data trends.
- A much-lowered mean time to resolution (MTTR) metric.
- It can automate any time-consuming process that is experienced by the threat hunting team.

Conducting the Actual Threat Hunting Exercise

Now that you have conducted a threat hunting risk assessment and picked a known model or framework that you would like to make use of, the next step is to actually launch the threat hunting exercise. It is made up of four distinct components, which are as follows:

1. Creating the hypothesis:
   In this first stage, you are trying to determine the results that you are expecting from conducting a hunting exercise. For example, if you want to penetrate the SQL Server Database, what are the statistical odds of it actually happening?

2. Collecting the relevant information and data:

Next, you will want to follow up with your hypothesis by investigating it with the threat hunting model or framework that you have selected. Some things to be on the lookout for include any anomalies, and malicious patterns in the datasets that have been compiled.

3. Determining the steps that can be completed by automation:

Once you have completed the last step, you then need to determine which steps can be carried out by automated threat hunting tools, such as using generative AI, machine learning, or even neural networks.

4. Plan a course of action:

After you have compiled the relevant information and data in the second step, the next step is to determine what happens next. If the analyses of the patterns and anomalies indicate that a cyberattack is in its beginning stages, then this needs to be handed over to the incident response team.

These components are illustrated in Figure 1.3

Although a significant amount of this chapter has been spent on threat hunting and penetration testing, these are the ultimate tools that can be used to find and remediate any gaps or weaknesses that are in existence from within your digital assets and physical assets. But it is also important to keep in mind that just because you and your IT security team have taken very proactive steps, this does not necessarily mean that you will be immune from becoming a victim of a cyberattack, or a similar security breach.

The truth of the matter is that any individual or business entity is prone to a cyberattack. What we have detailed in this chapter only provides the tools that will actually mitigate that risk from actually happening. So now the next question that you and your IT security team need to ask is: "What if we are hit with a cyberattack or a security breach? What do we do next?"

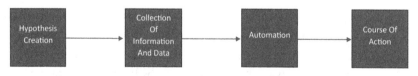

Figure 1.3

Obviously, the first thing that you would need to do is contain the breach by initiating your incident response plan. Then, you will need to restore mission-critical operations back to a normal state by launching your disaster recovery plan and your business continuity plan. Finally, the last step in this process is to determine how the cyberattack and/or security breach actually happened, and this is done through the use of what is known as "digital forensics". This will now be reviewed in the remainder of this chapter.

An Introduction to Digital Forensics

As it was just reviewed in the last section if your business has been hit with a security breach, and after you have more or less recovered and restored the mission-critical operations, the next step is to finally figure out how the cyberattack actually happened, and perhaps even try to discover who the perpetrator is. For example, was it a threat variant launched by a nation-state threat actor, or was it something that happened domestically? Given the digital world that we live in today, the source of the cyberattacker could emanate from anywhere in the world, and take place in just a matter of seconds.

To figure all of this out, the field of "forensics" becomes especially important. This is where specially trained investigators comb through a gargantuan amount of information and data in order to find the key pieces to not only determine what has happened but perhaps even bring the perpetrator to justice. But if they are from a nation-state threat actor, this could prove difficult to accomplish, because there will be extradition factors to take into consideration. From a technical perspective, forensics can be defined as follows:

> Forensic science is the use of scientific methods or expertise to investigate crimes or examine evidence that might be presented in a court of law. Forensic science comprises a diverse array of disciplines, from fingerprint and DNA analysis to anthropology and wildlife forensics.
>
> *https://www.nist.gov/forensic-science*

It should be noted that in many cases, forensics is also referred to as "forensic science", as you can see in the definition above. In most cases, this branch of forensics is often used in collecting evidence that can be found at a crime scene that is physical in nature. For example, in the case of a murder, forensics investigators will be primarily concerned with collecting any blood left at the scene, clothing, and perhaps even collecting biological samples from the victim themselves in order to confirm his or her identity.

But, another key area of prime interest for the forensics investigators will be collecting any DNA samples that are possible. This could be used not only in further identifying the victim with 100% accuracy but to even tracking down the person who was responsible for committing the murder. Once the perpetrator has been located, this evidence can then be used in a court of law in order to bring the perpetrator to justice. But once again, this is an example of forensics being used at a physical crime scene.

Now in the case of a cyberattack, this is where an entirely new branch of forensics will come into play. This is now called "digital forensics", because it has been the assets of a business of a digital nature that have been impacted. Therefore, the technical definition of it is as follows:

> Digital forensics is a branch of forensic science that focuses on identifying, acquiring, processing, analyzing, and reporting on data stored electronically. The main goal of digital forensics is to extract data from the electronic evidence, process it into actionable intelligence and present the findings for prosecution. All processes utilize sound forensic techniques to ensure the findings are admissible in court.
>
> *https://www.interpol.int/en/How-we-work/Innovation/Digital-forensics*

So as one can see from the above definition, the area of digital forensics is just as broad or even broader than the physical field of forensic science. It can also be a lot more complicated because there are many ways in which digital evidence can be collected, stored, and processed. For example, it touches on the following aspects of any IT/network infrastructure:

- Workstations
- Servers
- The entire network infrastructure

- Endpoints (the point of origination to the point of destination, and vice versa)
- Wireless devices
- Laptops
- Notebooks
- Tablets
- Smartphones
- Web applications
- Source code
- Anything that is made use of and stored in a cloud deployment model, most notably AWS and Microsoft Azure. It is important to note here that everything that resides in your AWS or Azure subscription is not only a target for the cyberattacker, but it can also be used for the purposes of a digital forensics investigation.

An illustration of digital forensics is shown in Figure 1.4.

A Brief History of Digital Forensics

At this point, it is important to provide a brief historical background into the field of digital forensics, in order to show the gravity of its importance in today's times. Here is a chronological history of it:

Figure 1.4

(From https://www.shutterstock.com/image-illustration/fingerprint-scanning-technology-concept-2d-illustration-1832462110)

- **Before the 1970s:** Digital forensics was not heard of, and very few people even had an understanding of it.
- **In the 1970s and 1980s:** Interest in digital forensics started to increase, given the advancements that were made in computer technology. However, at this time, the digital forensics Investigators were primarily from the Federal Government but had a computer background.
- **In the 1980s:** The first major area of digital forensics to be born was data storage, and it still continues this trend even today.
- **In 1984:** The FBI launched what is known as the "The Magnet Media Program". This was the first truly defined digital forensics program to be born.
- **In 1986:** Cliff Stoll, a Unix System Administrator at Lawrence Berkeley launched the first "honeypot" as the first true bait to trap cyberattackers and to capture their "digital footprints".
- **From 2003 to 2011 and 2001 to 2021:** This is the period when the simultaneous wars in both Iraq and Afghanistan took place. This led to the first true efforts in the international realm of digital forensics, as nations around the world started to share information, data, and other types and kinds of intelligence to mitigate the risks of terrorist attacks. Also, during this timeframe, the first set of best practices and standards came out for the field of digital forensics. The exact text of this can be seen at the link below:

    http://cyberresources.solutions/Digital_Forensics_Book/Best_Practices.pdf
- **To present:** Digital forensics is now being used on a daily basis in today's cyber threat landscape, especially as the threat variants get more sophisticated, covert, and stealthier. A prime example of this is ransomware.

The Methodology of Digital Forensics

A digital forensics exam can vary quite a bit on the specific use case. For example, a phishing attack may entail a different kind of collection and analysis, whereas a data exfiltration attack may require a completely different set of collection and analysis. But whatever the incident may be, digital forensics follows a common, distinct method-ology which is as follows:

1. The investigation:
   This is the first phase in which the impacted digital assets are identified, quarantined, and isolated.
2. The preservation:
   This is where the impacted digital assets are preserved and where "… Investigators copy evidential data, otherwise known as a 'forensic image'" (https://www.redpointcyber.com/what-is-digital-forensics-in-cyber-security/).
3. The analysis:
   This is where the digital forensics investigators examine col-lect and examine in great detail the pieces of evidence in the impacted digital assets.
4. The documentation:
   This is where the digital forensics investigators record in a written format all of the activities that they have engaged in. This is probably one of the most important components of this methodology, as the court of law will ask for this in order to confirm that the evidence has indeed remained intact, and thus admissible in a possible trial.
5. The presentation:
   This is where all of the documentation compiled in the last step is finalized into a final report, and used for presentation for all of the key stakeholders that have been involved in the digital forensics investigation.

This is illustrated in Figure 1.5.

The Benefits of the Digital Forensics Methodology

The above methodology brings certain benefits, which are as follows:

1. Incident response:
   By knowing how the security breach or the cyberattack hap-pened in the first place, the impacted business will be able to better respond to future threat variants.

Figure 1.5

2. Early threat detection:
   By putting into practice the lessons that have been learned, a business will now take proactive steps to implement generative AI tools that will provide alerts and warnings on a real-time basis.
3. Vulnerability management:
   As a result of the security breach or cyberattack, the impacted business will also take proactive steps to make sure that all vulnerabilities, gaps, backdoors, and weaknesses have been detected and remediated. This is where penetration testing and threat hunting will become essential, as reviewed in detail earlier in this book.
4. Malware analysis and preservation:
   Once the pieces of evidence have been carefully preserved, they can be used for further analysis by the CISO and their IT security team to take what has been learned and try to predict and extrapolate what future threat variants could look like down the road.

The Challenges of the Digital Forensics Methodology

Despite the advancements that the field of digital forensics has made, there are a number of key disadvantages to it. These are as follows:

1. The pace in technology:
   Simply put, the methodologies that have been incorporated into digital forensics simply cannot keep up with the pace of technological innovation that is happening today. A prime example of this is generative AI. There is strong interest now to use it in order to speed up investigations after a security breach or cyberattack has happened, but the procedures that are called for in digital forensics require meticulous detail and attention, which can take a large amount of time. In this regard, it is very important that evidence remains solid so that it can be admissible in a court of law.
2. The use of encryption:
   In order to keep documentation and other related information and data secure, forensics investigators now use encryption. As a result of this, there is now a great fear that transforming

these key pieces of evidence into a garbled state could affect their integrity and quality.

3. The volume:

    As the total number of cyberattacks increases in size, stealthiness, and covertness, this simply means that there will be that much more data that digital forensics investigators will have to parse and analyze through.

4. Issues with the cloud:

    In order to keep the evidence and other related information and data safe, many digital forensics teams are now making use of cloud-based deployments in order to store them in a safe and secure manner. While there is nothing inherently wrong with this, all of this could be stored in environments across different countries. In the end, the issue of which laws regulate the access and use of the evidence and the information/data becomes a key concern. For example, the laws in the United States will vary greatly from the laws in the EU, if cloud environments are used for storage there as well.

5. The cyberattacker:

    As just mentioned, while advancements are taking place in the field of digital forensics, the cyberattacker is usually one step ahead of this. For example, if a new forensics testing technique has been devised, it is quite likely that in a short period of time, the cyberattacker will have found a way to circumvent and penetrate it. As a result, any efforts to correlate the evidence with them will prove to be that much more difficult in the end.

The Analyses That the Digital Forensics Methodology Can Be Used For

As also mentioned earlier in this chapter, the field of digital forensics is a very large one. But in the end, there are ten major types of digital forensics analyses that can take place. These are as follows:

1. Computer forensics:

    This involves the study and investigation of computers, laptops, and storage computing devices.

2. Mobile device forensics:
   This involves the study and investigation of smartphones, SIM cards, mobile phones, GPS devices, tablets, PDAs, and game consoles.
3. Network forensics:
   This involves the study and investigation of any network activity.
4. Data analysis:
   This involves the study and investigation of both structured and unstructured data, as well as Big Data.
5. Database forensics:
   This involves the study and investigation of any access to a database and any changes made to the data.
6. Email forensics:
   This involves the study and investigation of the senders' and receivers' identities, the content of the messages, time stamps, sources, and metadata.
7. Malware forensics:
   This involves the study and investigation of different malware types to trace suspects and reasons for the attack and to determine the source code of the malware.
8. Memory forensics:
   This involves the study and investigation of the data from RAM. This is also technically referred to as "live acquisition".
9. Wireless forensics:
   This involves the study and investigation to analyze and investigate traffic in a wireless environment.
10. Disk forensics:
    This involves the study and investigation of hard drives and other physical storage devices, such as memory cards, servers, flash drives, and external USB sticks.

The Tools That Are Used in Digital Forensics

It should be noted at this point that when it comes to digital forensics, most of the tools that are used to conduct an investigation are software based. The primary reason for this is that most of the security breaches and cyberattacks that occur are targeted mostly to digital assets. Here is a sampling of some of the top software-based tools that are used in digital forensics:

1. Autopsy:
   This tool can be used for file filtering all the way to registry analysis of an OS, such as Windows 10/11, Linux, macOS, iOS, Android, etc.
2. FTK:
   This is an acronym that stands for "forensics toolkit". It can process massive amounts of data quickly, allowing for prompt analysis.
3. VIP 2.0:
   This is an acronym that stands for "video investigation portable". Its functionalities include everything from video retrieval to forensic reporting, thus making it a comprehensive solution.
4. Sleuth Kit:
   This software package is used to investigate a variety of file systems, and it ensures compatibility with a diverse range of file platforms.
5. Cellebrite UFD:
   This is a mobile forensics software that is used for data acquisition and analysis, and it supports a vast array of mobile devices.
6. X-Ways Forensics:
   This software package is used to investigate file structure analysis for data recovery.
7. Volatility:
   This is a software package that specializes in investigating and analyzing RAM memory dumps.
8. Magnet Axiom:
   This is a software package that specializes in acquiring data to visualizing connections.
9. The OS forensics:
   This is a software package that specializes in the sifting of vast amounts of data efficiently. It can also be used for password recovery which can be invaluable in investigations.
10. Paladin forensic suite:
    This is a software package that specializes in the imaging to analysis of images and videos.

The Differences Between Digital Forensics and Cyber Incident Response

Today, there is a lot of confusion between these digital forensics and cyber incident response. While the common denominator between the two is responding to a cyberattack and/or a security breach, they have different purposes entirely. Table 1.2 highlights these key differences between these two areas.

Overall, this chapter has reviewed the importance of doing a risk assessment analysis, and also how penetration testing and threat hunting are important in digital forensics. Also, an extensive review of both was provided as well. The chapter then provided an overview

Table 1.2

MILESTONE	DIGITAL FORENSICS	CYBER INCIDENT RESPONSE
Objective	It involves the investigation of all of the details of a cyberattack and/or security breach which involves: • How it occurred • The perpetrator(s) • What was compromised	This is the immediate response and action after a cyberattack and/or security breach is detected.
Methodology	It makes use of a variety of tools, techniques, and software packages to uncover digital evidence and to do reverse engineering.	Plans are used to manage communications and address regulatory compliance issues.
Timing	This happens after a cyberattack and/or security breach incident has been detected and there's a need to understand the source of it.	This is initiated as soon as a cyberattack and/or security breach is detected in order to minimize damage, loss, and downtime.
Resultant	Detailed reports are compiled that include: • How the incident occurred. • The extent of the damage. • Recommendations for prevention.	Detailed reports are compiled that include: • How the cyberattack and/or security breach incident actually occurred. • Documentation of the breach • Recommendations for future prevention and the lessons that have been learned.
Prevention	The main emphasis is on understanding past incidents, and evidence collection. The methodologies will also identify vulnerabilities that could lead to future breaches.	This incident response plan ensures that rapid containment of the threat variant happens. Also, its other main focus is on managing and mitigating active incidents.

of what digital forensics is all about. As it has been reviewed numerous times throughout this chapter, digital forensics is a very broad field, and it can be used to investigate just about any cyberattack and/or security breach that has impacted a digital asset that resides within the IT and network infrastructure. There can be many books written just on digital forensics alone.

But for purposes of this book, we will be focusing on how it relates to a cyberattack and/or security breach as it relates to data exfiltration, focusing primarily on a SQL Server Database. In the next chapter, we will take a closer look as to how digital forensics plays a role in this, and some of the methods that are used to break into a SQL Server Database. The chapter will then examine a specific attack that can be used to penetrate this kind of database, which is known as a "SQL Injection Attack".

The chapter after that will then examine how generative AI can be used in digital forensics as a means to collect evidence and the associated information and data.

2

DIGITAL FORENSICS, DATA LOSS, AND THE SQL SERVER DATABASE

Our last chapter dealt with the importance of conducting a risk assessment analysis, and also did a deep dive into what penetration testing and threat hunting are all about. Further, it was also discussed how important they are when it comes to the realm of digital forensics. Also, an overview was provided of what digital forensics is all about. It is important to keep in mind that digital forensics is a very broad field, and the methodologies that reside from within it can be used on any type or kind of digital asset that has been impacted by a cyberattack and/or security breach.

With that in mind, this chapter (and the next) will primarily focus on the database. In the examples that we show, we will be making use of the SQL Server Database. Therefore, this chapter will be divided into the following sections:

- Digital forensics as it relates to the SQL Server Database.
- An overview of SQL Server and some of its vulnerabilities (Chapter 3 will be exclusively devoted to what are known as "SQL Injection Attacks").

Digital Forensics as It Relates to the SQL Server Database

The Disk Drive

One of the first items that a digital forensics Investigator will examine is the hard disk of the impacted device, whether it is an actual

DOI: 10.1201/9781003469438-2

piece of hardware or virtualized. Therefore, it is important to examine what the components of a hard disk actually are. They are as follows:

1. The head:
   This is actually the physical element that does the reading and writing of the information and data onto magnetic material that is located directly into the hard drive, and most of them today have two "heads" per platter.
2. The track:
   Once the reading and writing of the information and data has been done, it then gets stored onto what is known as the "track" of the platter. These "tracks" are also magnetic and concentric by design.
3. The cylinder:
   These are when multiple platters are literally stacked upon one another, the tracks that reside in them interface with each other.
4. The sector:
   This is deemed to be the smallest unit that is available for storage in the disk drive, whether it is physical or virtual. The sectors are actually derived from the "tracks", as just reviewed. It should be noted here that each "sector" can contain up to 512 bytes of information and data.

The storage space on a hard disk can vary greatly, depending upon what kinds of requirements you have. If you do not offhand what the actual size of your hard drive is, here is a mathematical formula that you can make use of:

$$\text{Hard Drive Size} = \text{The Number of Platters} \times \text{The Number of Heads}$$
$$\times \text{The Number of Sectors} \times 512$$

The BIOS

The one key advantage that hard disks have today is that there is no need to manually configure them. This is all done through what is known as the "BIOS". This is an acronym that stands for "basic input/output system". It can be technically defined as follows:

> A basic input/output system or BIOS is a program fixed and embedded on a device's microprocessor that helps to initialize hardware operations and manage the data flow to and from the operating system (OS) at the time of bootup.
>
> *https://www.spiceworks.com/tech/devops/articles/what-is-bios/*

In turn, since the BIOS is such a critical component of the hard disk, it has its own set of functionalities. These are as follows:

1. The CMOS:
   This is an acronym that stands for "complementary metal-oxide semiconductors". This setup application allows the end user to modify hardware and system settings. The CMOS also refers to the nonvolatile memory of the BIOS.
2. The bootstrap loader:
   This is what locates the type and kind of operating system (OS) a device (whether physical or virtual) has when it is first started up.
3. The drivers:
   This is what identifies the device drivers and software that interact with the OS once it is fully initiated.
4. The POST:
   This is an acronym that stands for the "power-on self-test". This functionality confirms the computer's hardware before loading the OS.

An image of a hard drive is illustrated in Figure 2.1:

The Software

Now that we have reviewed the hardware aspects of what a digital forensics investigator focuses on, it is now important to review the software aspects as well. For this, we focus on the Windows OS that has been developed by Microsoft over the past few decades. The components that reside here are common between all of the different

Figure 2.1   An example of a hard drive.

(From https://www.shutterstock.com/image-photo/hard-disk-drive-platter-computer-mirror-1607332879)

flavors of the Windows OS, ranging all the way from Windows 3.1 to the present-day versions of both Windows 10 and 11. The following are the major components:

1. The cluster:
   These are actual groupings of "sectors". The primary intention here is to keep track of all of the files that have been stored from within the Windows OS. Of course, a larger hard disk (in terms of the storage of the information and data) will mean more "sectors" will be needed, thus requiring more "clusters".

2. The partition:
   This is also technically referred to as the "logical volume". These are actually smaller "logical units" that are used to keep the Windows OS working at a peak level of optimization. A huge risk here is if you see a large space on your hard disk that has not been divided, or "partitioned" up. This often means that the end user is storing information and data that are against the security policy of your business.

3. The MBR:

This is an acronym that stands for the "master boot record". Its main purpose is to hold the information and data about all of the "partitions" that have been formed on the actual hard drive (physical or virtual). It can also be used to keep track of any external storage devices that have been inserted into the device, such as USB drives.

The File Systems

From within the Windows OS, there are two types of file systems:

• The FAT
• The NTFS

These will be reviewed in the next two subsections.

The FAT

This is an acronym that stands for "file allocation table". This was designed to be used primarily with the floppy disk drives when the personal computer first came out, with Windows 3.1. It can be technically defined as follows:

> A file allocation table (FAT) is a file system developed for hard drives that originally used 12 or 16 bits for each cluster entry into the file allocation table. It is used by the operating system (OS) to manage files on hard drives and other computer systems.
>
> *https://www.techopedia.com/definition/1369/file-allocation-table-fat*

The FAT is also used for containing information and data about the files, time/date stamps, the directory structures and their corresponding names, and all of the metadata attributes that are stored from within the device (whether physical or virtual). There have been four different versions of FAT, and they are as follows:

1. The FAT 12:

This was only used on the floppy disk drives that were compatible with what is known as the "Microsoft Disk Operating System" (MSDOS). It could only store up 16 Mb of information and data.

2. The FAT 16:

This led the way from the floppy disk drive to the permanent hard disk. It could hold up to 2 Gb of information and data.

3. The FAT 32:

This file system has been used since Windows 95 first came out. It can store up to 2 Tb of information and data. It can still be used with older, but more recent versions of the Windows OS that have come out.

4. The VFAT:

This particular system consists of what is known as the "VxD driver". Interestingly enough, it was designed for file names longer than eight characters to be stored on the device (physical or virtual).

A unique feature of the FAT is that it has a functionality that is called "cluster chaining". This is where the end of one "cluster" points to and is linked to the next "cluster" in a chain. This chaining mechanism can work in any direction, but once that has been established, it cannot work in the other, or reverse direction.

The NTFS

This is an acronym that stands for "new technology file system". It can be technically defined as follows:

> NTFS, the primary file system for recent versions of Windows and Windows Server, provides a full set of features including security descriptors, encryption, disk quotas, and rich metadata.
>
> *https://learn.microsoft.com/en-us/windows-server/*
> *storage/file-server/ntfs-overview*

The NTFS has a number of key functionalities that include the following:

1. The enhanced file attributes:

This has the following capabilities:
- Read—only
- Archive

- System
- Hidden file attributes
- Granular rights, privileges, and permissions for all of the files, directories, and folders from within the device (physical or virtual)

2. The alternate data streams:

These are the various kinds and types of data structures that are attached to both existing and newly created files. These are typically viewed from the standpoint of what is known as "metadata", and it can be technically defined as follows:

Metadata contains information about a data asset, such as properties, origin, history, location, creation, ownership, and versions.

https://www.imperva.com/learn/data-security/metadata/

In other words, it is simply data about the data, and it is sometimes technically referred to as the "data fork".

3. The file compression:

This is a technique that is used to compress files of very large sizes and is commonly used today for images and videos that are simply too large to be stored. But when it comes to the NTFS, the compression algorithm that is used is called the "LZ77 file compression".

4. The encryption:

In regard to this, NTFS makes use of what is known as the "encryption file system". It makes use of the public key, which is used for encrypting at the point of origination, and for decrypting, the private key is used at the point of destination.

5. The journaling:

With this, the NTFS can actually keep a log of all of the changes made to the metadata, but not the actual data itself.

6. The shadow copy:

This is where the NTFS can literally take "snapshots" of all of the files and folders that reside in your device (physical or virtual). This can be used for a subsequent point in time in order to restore or rebuild a file or a folder.

7. The mount points:

This is a specialization where you can add more logical volumes without having to create more network drive letters. For example, in all of the Windows OSs, C:> is the default logical drive. Rather than creating an extra D:> or F:> logical drive, you use the mount point to add more logical volumes to what already exists.

The HFS

This is an acronym that stands for the "hierarchal file system" and is used most commonly across all macOS and iOS systems. It consists of the following functionalities:

1. The CNID:

This is an acronym that stands for "catalog node identification". This is a unique numerical value that is assigned to each and every file and folder within an Apple device.

2. The size:

This is the actual size of the folder or file and is located in the specified volume.

3. The time stamp:

This is the particular date/time when a folder and/or file was created, deleted, or modified, and even backed up.

4. The extent:

This is the specific area where the first part of any file or folder is stored upon the volume.

5. The fork:

This is the pointer to each and every file and folder in the Apple device, and these are stored on the volume.

It should be noted that the volume refers to the logical segmentation of the hard disk of a physical, Apple device. Also, a "catalog file" is used to store and update the version history of all of the files and folders that already exist, or are newly created.

The Linux and Unix

The file system here makes use of what are known as "blocks", and these represent the storage units in the device (physical or virtual). Also, all bits of information and data are considered to be individual files. The Linux and Unix OSs consist of the following characteristics with regard to their file system:

1. The boot block:
   This consists of specialized source code in order to initiate the launch of the OS when the device (physical or virtual) is first booted up.
2. The data block:
   These are the logical addresses that point to where each bit of data and information is stored on the device (physical or virtual).
3. The inode:
   This is a specialized file that points the block addresses to a file. The "inode" contains the following information:
   • The total number of bytes in the folder/file.
   • The timestamps.
   • The total number of blocks used by the file and/or directory.
   • The total number of links to a particular file.
   • The user and group ID numbers that correspond directly to the profiles that have been created.
4. The superblock:
   This functionality keeps track of and records all of the "inodes" and the status of all of the "blocks".

It is also important to note that it is the block size that directly determines how much information and data can be stored on the device (physical or virtual). The "block inode" also keeps track of any bad or damaged files due to a cyberattack and/or security breach, making it handy for a digital forensics investigation.

Retrieving Deleted Data

After a software exploitation attack, one of the prized possessions that a cyberattacker will go after is the datasets of the business. Typically, they would be most interested in stealing what are known as the "personal identifiable datasets" (PII). Very often, these datasets will be sold onto the dark web for a nice profit, or worst yet, they could be used in an extortion-style attack. But whatever the case, these kinds of security breaches are typically known as "data exfiltration". The technical definition of it is as follows:

> Data exfiltration is data theft, the intentional, unauthorized transfer of data from a system or network. Various agents target data exfiltration—attackers, insiders, and malware designed for data theft.

> *https://www.paloaltonetworks.com/cyberpedia/data-exfiltration*

The finding out of why and how data exfiltration actually happens is of prime interest to any digital forensics investigator. In this section, we review as to how data that have been lost, stolen, or even deleted can actually be recovered in the end.

The Deleted Files

It is very important to note at this point, that if data are simply deleted, it is permanently gone and can never be recovered. But, this is far both the truth and reality. For example, when a file is first deleted, the OS (as reviewed in detail in the last section) first puts a marker on it to let the file management system know that it is no longer in a specific cluster or block. This is technically known as a "logical deletion", but at this point, the data still remain in the hard disk. This is the crucial point at which digital forensics investigators begin their detective work.

Another key concept is what is known as the "unallocated space". This simply refers to the fact that there is a part of the hard drive that has not been portioned yet. If this is discovered, for example, by a rogue employee, they can then attempt to heist and store the PII datasets in this area. In fact, this is a very common tactic that is used, and this is also one of the very first areas that the digital forensics investigator will go after. Also, keep in mind that today, they have very sophisticated tools to find deleted or heisted data quickly and efficiently.

Gaining Back Cached Files

There is yet another area where information and data can be stored, and this is known as the "cache memory". It can be technically defined as:

> Cache memory is a small-sized type of volatile computer memory that provides high-speed data access to a processor and stores frequently used computer programs, applications and data.

> *https://www.techopedia.com/definition/6307/cache-memory*

So, for example, suppose there is a website that you visit quite often, such as your banking or credit card portal. Rather than having to process each request that you trigger every time, that website address (or the "URL") will be stored in the "cache memory" so that you can access it in seconds. If this does not exist, your device (physical or virtual) will have to process this request from scratch, thus consuming valuable computing and processing resources.

Figure 2.2 illustrates the cache memory chip.

Figure 2.2   An example of the cache memoey.

(From HTTPS://WWW.SHUTTERSTOCK.COM/IMAGE-PHOTO/CLOSE-VIEW-CENTRAL-PROCESSING-UNIT-CPU-2459555331)

So now the question is, how do you retrieve deleted or heisted data from a cache memory? Although the exact steps will vary on the circumstances and the situation, here is the general methodology that is used:

- Determine any keywords that have been associated with your file.
- Using the appropriate digital forensic software (as reviewed in Chapter 1), enter these keywords, and anything else that could be unique about the data that were deleted or heisted.
- Once the last step has been accomplished, you will then have to manually go through all of the results that have been presented to you, and from there ascertain which of the datasets you will need to retrieve. But of course, with the birth of generative AI, this process could also very well be automated in future digital forensics software packages.
- Next, have the software look for any data that could have been "disposed" in the unallocated spaces of the hard disk. Some key items that you will need to enter are the file headers and the file extension types (e.g., .DOC, .XLS, .PDF, .PPT, etc.).

Gaining Back Files in Slack Areas

There is yet another area on the hard disk known as the "slack space". It can be technically defined as follows:

> Slack space refers to the unused or wasted storage space within a file or on a storage medium. It occurs when the actual data stored in a file or on a disk does not completely fill up the allocated space. This unused space can be found at the end of a file or within sectors on a disk that are not used by any file.

> *https://www.lenovo.com/us/en/glossary/slack-space/*
> *?orgRef=https%253A%252F%252Fwww.google.com%252F*

In other words, imagine that you have partitioned out your entire hard disk. If there are areas that do not store any data, it becomes unused, also known as the "slack space". If the cyberattacker detects this, he or she can use it as a "backdoor" to access other areas of the hard drive and launch a data exfiltration attack. But if there are some data that reside here and are still deleted or heisted, you will want to follow this general methodology:

- Using the digital forensics software package, try to locate and ascertain any remnants or fragments of the data.
- Insert a "header" in one of these (or more) data remnants.
- Once the digital forensics software package has found all of the data that are associated with this particular header, you will then need to manually comb through the results to find what you are missing. It could also be the case that you will need to enter multiple headers to zero in on the deleted or heisted data that you are looking for.

The Nonaccessible Area

There is an area on your hard disk that will be deemed as "nonaccessible". From the standpoint of cybersecurity, it can be technically defined as:

Hard disk drive is not accessible means Windows prevent you from accessing the files and folders on the partition, and you're unable to modify, delete and add files on this partition.

https://www.diskpart.com/articles/drive-not-accessible-access-denied-5740i.html#:~:text=Hard%20disk%20drive%20is%20not%20 20accessible%20means%20Windows%20prevent%20you, add%20files%20on%20this%20partition

But another way of looking at it from the standpoint of digital forensics is to think of it as an area that is simply not recognizable to the OS, for whatever reason. However, it might not be accessible because some of the sectors in the "nonaccessible" could be damaged, or there may be certain limitations in the OS that restrict it from accessing this particular space.

But all is not lost here either, as the digital forensics Investigator can use what is known as a "hex editor" to try to retrieve any deleted or heisted data that could be in this area. A hex editor can be technically defined as follows:

> A hexadecimal (hex) editor (also called a binary file editor or byte editor) is a computer program you can use to manipulate the fundamental binary data that constitutes a computer file.

> *https://ctf101.org/forensics/what-is-a-hex-editor/*

The RAM and Digital Forensics

This is an acronym that stands for "random access memory". It is a core part of the device (physical or virtual) and is considered to be "volatile" in the sense that it will only store and retain data for as long as your device remains on. Once it is off, it will no longer retain the data. Thus, one of the primary goals of digital forensics in this regard is to avoid altering the data in any way, shape, or form.

But in order to do this, you first need to have what is known as a "forensics agent". With regard to the Windows OS, if you install this somewhere after the device (physical or virtual) is being used, there is a very high statistical probability that you will alter whatever data remain in the RAM as you search for the deleted or heisted one. So therefore, it is highly recommended that once you purchase a new device (physical or virtual), you take the time to install this tool before you start making use of it.

But on the other side, this issue does not exist with the Linux/Unix OSs. You execute and run various command line sequences so that the existing data in the RAM are not altered by any means.

The Registry

One of the key components of any OS, and in particular that of Windows, is what is known as the "registry". It can be technically defined as follows:

> The Windows registry is a centralized, hierarchical database that manages resources and stores configuration settings for applications on the Windows operating system.

> *https://www.avast.com/c-windows-registry*

Given its sheer level of importance, the Windows registry has not only become a prime target for the cyberattacker, but also it is one of the areas where the digital forensics investigator first starts their examination of a device (physical or virtual). Therefore, it is important to review its key components, which are as follows:

1. The password information:

   Although in Microsoft Azure, it is the Azure Active Directory that is used to create groups and profiles, it can also be used to store login credentials, such as usernames and passwords. However, the Windows registry can also store these credentials.

2. The startup:

   The Windows registry also contains the key packages that are needed to boot up the OS when the device (physical or virtual) is turned on.

3. The storage device hardware:

   The Windows registry also contains a database of all of the storage devices that are connected to the device (physical or virtual).

4. The wireless network:

   There is a database that collects all of the information and data for every network that the device (physical or virtual) connects to. In this instance, it is the "service set identifier" (SSID) that is primarily recorded. This is simply a unique set of characters that uniquely identifies the wireless network that you are connected to.

5. The internet information:

   Anything that you do on the internet is also recorded in the Windows registry. Such information and data that are collected include the URLs and the download path directory structure for anything that you may have downloaded.

6. The unread email:

   The Windows registry also keeps track of any email messages that have gone unread, and the timestamp information and data for every email that has been sent and received.

The Constraint to Search Filtering

As it was mentioned earlier in this chapter, a digital forensics investigation is not completely an automated process yet. In this regard, one of the manual mechanisms that is involved is in putting in keywords and other relevant search permutations into the digital forensics software package to locate the deleted or heisted data that you are looking for.

But in this regard, the primary constraint here is that businesses are literally overflowing with all kinds of data, both qualitative and quantitative. Much of this is brought upon the digital world that exists today, especially where all devices (physical or virtual) are interconnected with one another. The primary catalyst for this has been the "Internet of Things" (IoT).

So, the best way to proceed forward, in this case, is to only examine those areas of the hard disk where you think the data have been deleted or heisted. Taking this approach could also lead to other avenues of discovery, which may not have been suspected before.

How to Extract Forensics Data

It is important to note that a majority of the digital forensics software packages have become very sophisticated in nature, and it is highly expected that this trend will continue especially as generative AI will start to become incorporated into it. It is also highly likely that these specific software packages will even be installed in future releases of OSs by default, at each and every endpoint. These are typically the devices (physical or virtual), servers, databases, etc. As a result, digital forensics data will be automatically collected by the system, which can then easily interface with the software package that has been deployed into the OS.

This is illustrated in Figure 2.3.

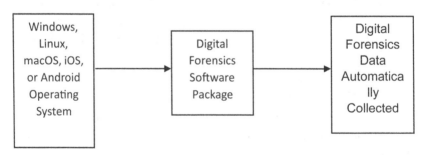

Figure 2.3 The OS process to digital forensics.

Collecting the Data

In order to collect the forensics data in this kind of example, it is important for you to follow these steps:

1. Acquire an image of the data that you are intending to collect and analyze. This should have all of the contents of the hard disk in that particular partition. From here, then sort the data by file type that has either been deleted or heisted.
2. Once you have found the particular file you are interested in investigating, it is important now to carefully examine the metadata that is contained within it. As it was reviewed earlier in this book, the metadata is simply the granular levels of data that describe the overall, macro view of the data. It is also very important at this stage to examine other items such as the date/time stamps and any related file headers.
3. Once you have done the last step, it is now important to test the contents in the file structure that you have chosen to further investigate. Depending upon the digital forensics software package that you are making use of, the results of this will typically appear in a graphical format. In this way, you will be able to discern quickly if the internal structure actually matches the file header (as reviewed in the last step).
4. Retain the files:
   At this point, it is very important to save everything that you are investigating. There are two primary purposes for this:
   - So it can be used as evidence in any litigation that comes forth at a court of law.
   - It is critical that this be included in a digital forensics report that will be accessed by the key stakeholders who are involved in the investigation. Although the exact format of the report will vary from investigation to investigation, the following is typically included:
     – The title:
       This includes the name of the investigation, the members of the digital forensics team, and the dates that the digital forensics investigation actually took place.

- The table of contents:
  Just like in any piece of documentation, there should be a table of contents.
- The case summary:
  This is considered to be the "executive summary" of the digital forensics report, and it should only be at most a few paragraphs long, reviewing the highlights of the digital forensics investigation.
- The evidence:
  In this section of the digital forensics report, all of the evidence that has been collected and analyzed needs to be both listed and described in detail. Include screenshots if needed as well.
- The objectives:
  This will detail why the digital forensics investigation took place, and the reasons for it should be detailed.
- The actual investigation:
  This section should outline the procedures that were taken during the digital forensics investigation, including the digital forensics methodology that was used. It should also have additional explanations that extrapolate the evidence and how the very important conclusions were formulated.

  The chain of custody also needs to be included in this section. It can be technically defined as follows:

> The chain of custody is the paper trail or chronological documentation of digital evidence. It establishes the sequence of control and the process for collecting, transferring, and analyzing the data in question. It keeps a record of every person who handled the evidence, the dates and times the data was collected or transferred, and the purpose for the transfer. If any information is missing from the chain of custody, the evidence could become inadmissible in court.
>
> *https://cornerstonediscovery.com/maintaining-chain-of-custody-in-digital-forensics-what-you-should-know/#:~:text=The%20chain%20of%20custody%20is,analyzing%20the%20data%20in%20question*

To view an established framework for writing a digital forensics report, access the following link: http://cyberresources. solutions/Digital_Forensics_Book/CISA_FRAMEWORK. pdf

- The tools used:
  In this section, you need to mention all of the digital forensics tools that were used, both the hardware and/or the software. Also, including the relevant screenshots of them will help make it sound clearer and more understandable.
- The findings:
  This is where you have to review and explain each piece of evidence and clearly demonstrate what it proves. At this point, it is very important to reiterate the most important conclusions.
- The next steps:
  In this section, you propose what the follow-up should be. Most likely, it will be about how the evidence will be used in a court of law.
- The appendix:
  In this section, you include any supporting documentation to further corroborate the evidence and conclusions that you have formulated.
- The glossary:
  At this point, you will also want to include a breakdown of the key, technical terms that have been used. This way, all of the key stakeholders that were involved in digital forensics will be able to understand them, especially those who do not have a digital forensics background.

5. Extraction:
   At this stage, you will extract all of the information and data from the recovered data, but most importantly, this should also include any hidden data and metadata.
6. Make copies:
   At this phase, you will want to make duplicate copies of everything that was collected and analyzed. This serves two purposes:
   - Backup and recovery
   - Archiving (many of the data privacy laws mandate this, such as the GDPR, CCPA, HIPAA, etc.)

7. The reconstruction:

In this last stage, which is of course the most difficult for any digital forensics investigator, you will need to reconstruct the recovered. This is probably the most time-consuming and laborious aspect of this entire process.

The SQL Server Database

The last section of this chapter provided a detailed overview of how digital forensics can be used to investigate data that have been deleted or even stolen. In this section, we now review how the hacks can be done in an actual database. Most of us have probably heard of what a database actually is. It can be technically defined as follows:

A database is an organized collection of structured information, or data, typically stored electronically in a computer system. A database is usually controlled by a database management system (DBMS). Together, the data and the DBMS, along with the applications that are associated with them, are referred to as a database system, often shortened to just database.

https://www.oracle.com/database/what-is-database/

Although this is one of the simplest versions of a database, in today's world, they are actually quite complex, and in many instances, they need an entire team just to work with it. There are many brands of databases and a sampling of them includes the following:

- Oracle
- Microsoft Access
- Microsoft SQL Server
- PostgreSQL
- MySQL

For purposes of this chapter, we will be primarily focusing on the SQL Server. This has been developed and launched by Microsoft, and the most recent version of it is SQL Server 2022. Many of these database packages, while still available through retail channels, are now mostly available through any cloud platform, such as AWS or Microsoft Azure. In these cases, you will actually get the "virtualized" version of the database. There are three key advantages to this:

- Cost savings: Rather than paying literally thousands of dollars for the retail version, you only pay a fraction of that when you get it through AWS or Microsoft Azure.
- The virtualized version of the database, as opposed to the retail version, can be deployed in just a matter of a few minutes, thus allowing teams to meet project deadlines on time.
- By creating a virtualized database, you will get full technical support and the needed software patches and upgrades will be deployed automatically for you.

Defining the SQL Server Database

At this point, it is important to have a broad and general understanding of what a SQL Server database is all about. Microsoft SQL Server is what is known as a "relational database management system" (RDBMS). It has many sophisticated functionalities, which support any activity that requires a transaction to take place. A prime example of this is an e-commerce store. There are many steps that are involved in selecting a product or service, and the final transaction that occurs is the actual purchase for them. Also, given that generative AI is now starting to be deployed in just about every application imaginable, SQL Server can now support this as well.

The root of the SQL Server is the "Transact-SQL" (T-SQL). This is a proprietary database programming language that has been developed by Microsoft. The following are the major components of the SQL Server database:

- It has a row-based table structure.
- It can connect different datasets that exist in different tables to one another.
- It makes use of what is known as "referential integrity", to make sure that the database is fully optimized at all times.
- The database engine, which controls and processes all of the transactions that take place on a SQL Server database.

In fact, the overall package can support up to 50 different database engines.

- It consists of stored procedures and triggers, which are also launched by the database engine.

In order to connect to a SQL Server database, you will need to do so either via what is known as the "command line" or "graphical user interface" (GUI). There are other key pieces of information that are required, and they are as follows:

- The instance name of the database engine.
- The specific network protocol and the connection port that is being used to connect to the SQL Server database.

There are still other components to the database engine, which are as follows:

- The main purpose of SQL Server OS (SQLOS) is to process the lower level functions, which are:
  - Memory optimization.
  - Input/output (I/O) management.
  - The job scheduling of the transactions that are to take place.
  - The locking of the datasets to avoid any conflicts when they are being processed.
  - A "network interface layer" (NIL) that resides above the database engine. This is used to connect with the other SQL Server databases, which will most likely be hosted on other different servers.

The Other Versions of SQL Server

As it has been mentioned previously, the SQL Server is typically used in the relational database management system. But there are also other flavors that can appear, although they are not widely used exclusively for the SQL Server database. The following are the different versions:

1. The hierarchical database management system:
   This type of SQL Server is used to manage those kinds of datasets that are organized in a tree-like structure. Examples of this include:
   - The parent-child relationship
   - The top-down

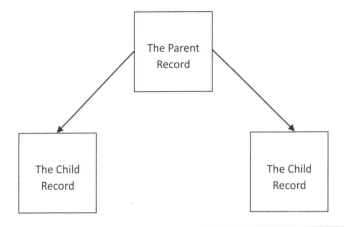

Figure 2.4 Records in a SQL database.

- The bottom-up

    It is important to note here that the first record is known as the "root or parent record". Beneath this are the "child records". This is illustrated in Figure 2.4.

    The primary advantage of using this kind of SQL Server model is that it is easy to add and/or remove datasets because the relationships between them have been defined and established.

2. The network database management system:

    In this kind of SQL Server model, each child record can have more than one parent record, thus allowing the team to create much more complex relationships between the datasets. Mathematically, this is also known as a "many-to-many relationship" or "N:N". This is illustrated in Figure 2.5.

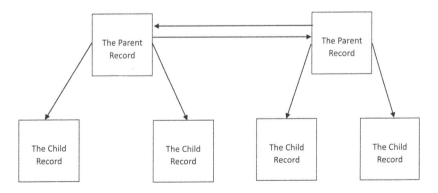

Figure 2.5 A many to many relationship in a SQL database.

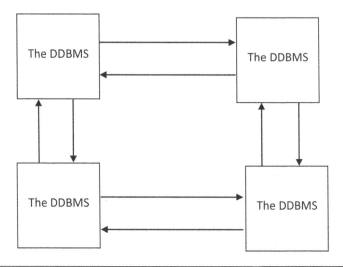

Figure 2.6   Access of datasets in a SQL database.

Also, the datasets are created into a graph structure and accessed through various "paths".

3. The distributed database management system:
   This is also known as the "DDBMS". This kind of SQL Server model is actually a collection of other distributed databases. This kind of model allows for the synchronization of the datasets so that they can be shared. This all happens in a transparent and seamless fashion. This is illustrated in Figure 2.6.

4. The object-oriented database management system:
   This is also known as the "OODBMS". In this regard, all of the datasets and associated information are represented in the form of various "objects". As a result, this does not allow for the relational rows to be mapped to the relational columns, and vice versa. However, the primary advantage here is that complex datasets (both quantitative and qualitative) can be contained in their properties and metadata. Also, it is possible to have many-to-many ("N:N") relationships that can be accessed through what are known as "pointers".

5. The multi-model database management system:
   This is also known as the "MMDMS". It is typically used to support more than one kind or type of dataset. The key advantage here is that the team is not limited to only one

database model. Rather, by making use of the "MMDMS", you gain much more flexibility in working with different datasets. Also, this kind of SQL Server model can support relational and wide-column database schemas.

6. The cloud database management system:
This is also known as the "CDMS". With this, the entire SQL Server model is deployed and hosted as a "cloud service software system". In other words, all of the networking, hardware, and software are virtualized onto a cloud-based platform, as described earlier in this chapter. As a result, all of the data are stored and managed in a secure manner in either AWS or Microsoft Azure and thus can be accessed from anywhere at any time, provided that there is a solid internet connection. One of the key advantages to this is that the SQL Server model becomes highly scalable, and thus, entirely eliminates the need for an on-premises deployment. Because of this, all of the costs of hosting are primarily based upon the consumption of the resources that are used to host the SQL Server model. Also, the team can make further use of what is known as a "mixed cloud environment". This is where a combination of private, public, or hybrid cloud-based deployments comes into play.

7. The real-time database management system:
This is also known as the "RTDMS". This kind of SQL Server model makes use of real-time processing in order to keep up with and effectively manage the dynamic datasets. For example, the datasets are processed immediately when they are received by the database. This is in sharp contrast to the older SQL Server models where "persistent data" is usually stored. This can be technically defined as follows:

> Persistent data is any data stored on a non-volatile storage medium that remains accessible for long-term use until it's purposefully deleted or overwritten.
>
> *https://www.purestorage.com/knowledge/*
> *what-is-persistent-data.html*

Examples of the type of industries where this kind of SQL Server model would be deployed and used are the supply chain/logistics and the financial markets. The primary reason for this is that the data are collected very quickly (and lots of it), and because of the needs of the end user, they must be processed right away, without delay.

The Application of the SQL Server

There are many kinds and types of applications that a SQL Server database can be used for. For example, it could be used to drive an e-commerce store, as customers select products and services, and makes payment for them, or it could be simply used to house a gargantuan amount of information and data (which is technically known as "Big Data") that will be used at some subsequent point in time, such as developing competitive intelligence or conducting market research.

Probably the most common use of a SQL Server database is in what is known as a "web-based application". A technical definition of this is as follows:

A web application is software that runs in your web browser. Businesses have to exchange information and deliver services remotely. They use web applications to connect with customers conveniently and securely. The most common website features like shopping carts, product search and filtering, instant messaging, and social media newsfeeds …

https://aws.amazon.com/what-is/web-application/

As a real-world example, suppose you visit the website of a giant retailer, like Walmart. They will of course have an online store, in which you can select the products that you want to acquire. Once you have done this, you make the payment, which gets processed, and then the products get delivered to your doorstep. In other words, you are accessing a service that is deployed remotely, in order to fulfill the needs of your daily lifestyle.

While all of this may seem simple on the outset, the truth of the matter is that behind the scenes, it is a quite complex process that is

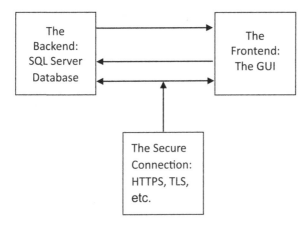

Figure 2.7 The secure connection between the backend and frontend of a SQL database.

taking place. In other words, there are two major components to the Walmart Store. The "back end", which is the SQL Server database, and the "front end", which is the GUI that you interact with in order to select your products and make your payment. This is illustrated in Figure 2.7.

In this regard, it is the SQL Server database that is probably the most important component. For example, not only does it store a history of your product selections, but it also stores your financial details so that you do not have to keep entering your credit card number all the time (due to cybersecurity issues, most online retailers of today give you that choice of whether to store that kind of information or not).

The Weaknesses of Web Applications

Despite all of the efforts that are taken to "harden" both the front end and the back end, vulnerabilities can still persist. This can not only pose a grave risk to the end user but can also have serious consequences for the SQL Server database, and all of the datasets that reside from within it. For example, although a SQL Injection Attack can take place directly on the database itself, the cyberattacker can also find a way to hack into the actual web application interface itself. Therefore, it is important to review at this point what some of these actual hacks are, because in the event of any data that have been deleted or heisted, this is one of the very first areas that the digital forensics investigator will go for an examination in an effort to determine what has truly happened. Some of the possible hacks in this regard include the following:

1. The directory traversal:

   This is actually deemed to be a very basic weakness from the start. In this type of attack, the cyberattacker simply looks up and browses for any kind of web application. From here, they could then very well have the ability to glean more information about the directory structure of the SQL Server database or the server that actually hosts the web application. If they are successful, then it is quite possible that they could see any files that are deemed to be "sensitive" in nature. One of the easiest ways to do this task is to make use of what is known as a "crawler" or "spider", a tool that is actually widely publicly available. This can be technically defined as follows:

   > A web crawler, crawler or web spider, is a computer program that's used to search and automatically index website content and other information over the internet.

   https://www.techtarget.com/whatis/definition/crawler

   For example, the "crawler" or "bot" is typically used by the major search engines, such as Google and Bing. While the intention here is not malicious, the goal is to comb through millions upon millions of websites in order to search for their particular keywords and rank them accordingly, whenever a query is actually submitted. In this regard, a publicly available tool is the "HTTrack Website Copier", which is available for download at: httrack.com. While this tool can glean a lot of information in a safe and secure manner, if a cyberattacker knows how to manipulate it the other way, a lot of sensitive information can be revealed, as discussed earlier. To mitigate this particular kind of threat to your SQL Server Database, it is highly recommended that you follow these tips:
   - Do not store any kind of sensitive files or data on the web server that is hosting your particular application. In other words, the only files that appear in these directory structures:
     - /htdocs
     - DocumentRoot

- Should be those that are required to make the web application to function the way it is supposed to.
- Reconfigure the "robots.txt" file so that something like the "HTTrack Website Copier" cannot probe for any sensitive or confidential files.
- Make sure that public access is only given to any other remaining folders that are needed to make your web application operate the way it is supposed to. The concept of least privilege must be strictly observed here. In order to do this, the following are directory structures that will need to be reconfigured:
  - For Linux/Unix-based OSs: .htaccess
  - For Windows OSs: httpd.conf
- Consider deploying a "honeypot". This can be technically defined as follows:

> A honeypot is a cybersecurity mechanism that uses a manufactured attack target to lure cybercriminals away from legitimate targets. They also gather intelligence about the identity, methods and motivations of adversaries.
>
> *https://www.crowdstrike.com/cybersecurity-101/*
> *honeypots-in-cybersecurity-explained/*

In other words, create a phony directory that contains fake data so that any "crawler" or "spider" will penetrate into it. The main benefit of this is that by sacrificing this kind of fake data, you will be able to see how these tools operate (in fact for that matter, even the cyberattacker), which will give a much better idea then as to how you need to further fortify those sensitive folders in your SQL Server Database or web application.

2. The buffer overflow:
   In the OSs of today, any information or data that is written to it is assigned a fixed length of space. A buffer overflow attack happens when this amount has been surpassed. A technical definition of this is as follows:

> Buffer overflows occur when the amount of data written to one of these blocks of memory exceeds its size. As a result, memory

allocated for other purposes is overwritten, which can have various effects on the program.

https://www.checkpoint.com/cyber-hub/cyber-security/
what-is-cyber-attack/what-is-a-buffer-overflow/

As a result, and according to the technical definition, a side effect of this kind of attack could spill over to the SQL Server Database, by adversely affecting the fixed size and allocation of its memory buffers. Some of the best ways to mitigate against this kind of threat include the following:

- Conducting regular scans of the source code of both the web application and SQL Server Database.
- Always apply the needed software patches to the web application and the SQL Server Database in a timely fashion.
- Always scan for weaknesses and gaps on a regular basis by making use of a reputable vulnerability scanner.

3. The URL manipulation:

With this kind of particular attack, an automated input is created that can manipulate the URL of a legitimate web application and send it back to the server. From here, the end user will most likely be directed to a fake or spoofed website, where they will be conned to give up their login credentials, or other types and kinds of confidential information and data. It should be noted that this kind of attack does not happen all the time, it is only periodic by nature as many of the domain registrars have taken steps to prevent any domains that are registered through them from being manipulated or tampered with. Another good way to mitigate this attack from happening to your SQL Server Database or web application is to make use of what is known as a "next-generation firewall", which actually makes use of generative AI and machine learning to intercept malicious or rogue-based data packets.

4. The hidden field impersonation:

In this particular scenario, a web application will actually have what are known as "hidden fields" that are actually embedded into the source code. An example of how this would look is as follows:

```
<input type= "hidden">
```

Because of this, any confidential information or data that the end user submits will be covertly stored here and, thus, pose a grave security risk. Some of the best ways to mitigate this kind of attack include the following:

- Make sure that you follow secure source code practices.
- Make use of the HTTPS protocol.
- Make use of Anti-CSRF tokens: This requires the use of a specialized token before any kind of information can be submitted on a "Contact Us" form.
- Deploy server-side validation: This can be used to confirm that the information and data that are submitted by the end user are actually valid based on defined permutations.
- Stay regular on the software update patching schedule.
- Make use of next-generation security tools, such as that of the next-generation firewall.

5. The cross-site scripting:

This is also commonly referred to as an "XSS attack". In this kind of situation, there is a vulnerability in either the SQL Server data or the web application itself (or perhaps both of them) that interacts with the end user input in a malicious manner. Typically, the main culprits are the weaknesses and gaps that exist in the compiled Java code. This kind of attack can also redirect an end user to a phony or spoofed website from which their login credentials can be covertly captured and used for nefarious purposes, such as launching an identity theft attack or an exploitation attack. Even worse, the XSS attack can allow cyberattacker to penetrate into the wireless device of the end user when they access a particular web application that makes use of JavaScript. From here, the hacker will then have full "read" and "write" access to the wireless device or manipulate the web browser cookies in a malicious way. Some key methods to mitigate this from happening include the following:

- Confirm and validate all end user input that you receive, especially those that are submitted on the "Contact Us" page.
- Sanitize all incoming data and information. A tool like "HTML sanitizer" will work very well here, and it can be accessed using the following link:

    https://github.com/mganss/HtmlSanitizer

- Never allow for the end user to enter any kind or type of HTML code onto the web application, once again, especially on the "Contact Us" page.
- Apply a set of best practices and standards when you deploy cookies on the end user's web browser.
- Also make use of a web application firewall. With this, you create specific rules and permutations that will forbid the entry of any kind or type of HTML code.
- Never display static values onto the end user's web browser.
- Filter out for and delete any kinds of "<script>" tags that might be present on the input fields in the "Contact Us" page.
- Only make use of web application and SQL Server Database error message as they are absolutely. Even these can reveal information and data about them.

6. The default script attacks:

This is a type of threat variant that makes use of popular source code scripting languages such as Perl and the hypertext processor (also known more commonly as "PHP"). A script can be technically defined as follows:

> A script or scripting language is a computer language with several commands within a file capable of being executed without being compiled.

> *https://www.computerhope.com/jargon/s/script.htm*

In this regard, there is often confusion as to what "source code" and "script" are. With the former, these are the entire modules of code that are compiled in order to create and launch a full web application. But the latter is merely a subcomponent of a module (in fact the script itself can even be considered to be a "submodule") that just does one small part in the overall functionality of the web application. But in the end, no matter what it does, the same security vulnerabilities also exist in a script as they would in an actual source code if the programming is not written properly. Probably the biggest issue here is that the software development team often leaves backdoors so that they can access the code quickly. At the time of the final delivery to the client, these backdoors are often forgotten, and thus they become a literal "way in" for the cyberattacker. So, the best way to avoid this is to test to make sure that all the backdoors have been deleted before the final delivery actually takes place. However, there are other techniques that one can use, and they are as follows:

- Make sure you and your team know the exact purpose of why you are creating a script in the first place. Then, take the time to fully understand it by thinking out how it will integrate into the overall web application.
- Make sure to delete any sample scripts from the servers that are hosting the actual SQL Server Database and the web application.
- If you do make use of what is known as a "content management system" (such as WordPress, Joomla, Drupal, etc.) make sure that you install and deploy all of the needed software patches and upgrades to them. Believe it or not, they also tend to be a prize target for the cyberattacker.

7. The insecure login mechanism:

If your web application is one that will make use of heavy end-user interaction, such as that of an online store, one of the functionalities that you will want to incorporate into is that of a customer portal. This is where the end user can create their own personal account, to view past purchases, pending orders, payment confirmations, etc. But in order to initiate this process, they will have to first create a user ID and associated password. This too will have to be implemented into the source code, and an area where software developers fail to recognize is the need for the web application to properly handle any login credentials that are incorrect. As a result, this could leave information for the cyberattacker to use to not only hack into the web application itself but to also even the SQL Server Database. Therefore, you should test the mechanism that you create before it is given to the client, and ultimately released into the production environment. This can be accomplished by using these three scenarios:

- Make use of an invalid user ID with a valid password.
- Make use of a valid user ID and invalid password.
- Make use of both an invalid user ID and an invalid password.

As for a further and more comprehensive testing approach, it is also wise to make use of a "password cracking" tool to make sure in the end that the login mechanism you have created is as hacker-proof as possible. Also, here are some other recommendations as well to make sure that your login mechanism is as safe as possible:

- Any login errors that you create (the end user will receive this if they enter an invalid user ID and/or password) should be a generic possible. The primary reason for this is that if you make it specific enough, it can actually give away information that can be used to launch a cyberattack.
- Never return an error code that returns a specific code when an invalid user ID and/or invalid password is used.
- Always make use of a "CAPTCHA" or something like a mathematical equation that will make it difficult for a password cracker to be used.

Two different examples of "CAPTCHAS" can be seen in Figures 2.8 and 2.9, respectively.

Figure 2.8 A CAPTCHA.

(From https://www.shutterstock.com/image-photo/captcha-on-laptop-screen-dubai-united-2413327199)

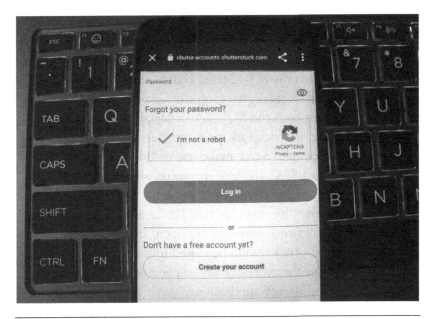

Figure 2.9 A CAPTCHA.

(From https://www.shutterstock.com/image-photo/indonesia-21-december-2023-filling-captcha-2404984335)

- Deploy an "intruder lockout mechanism" that will completely disable the login mechanism after a series of unsuccessful login attempts (usually three is the number of times that is used).
- To make your login mechanism even that much more robust in terms of security, implement the use of either two-factor authentication (2FA) or multi-factor authentication (MFA). Under both of these scenarios, at least two or more differing authentication mechanisms are used to fully confirm the identity of the end user.

Further Steps to Protect the Web Application and the SQL Server Database

In addition to all of the recommendations that have been reviewed so far in this chapter, here are some more suggestions that you should take into serious consideration as well:

- Use different servers to host the actual SQL Server Database and web application.

- If you are deploying both of them into a cloud environment, such as AWS or Microsoft Azure, make use of all of the security tools and functionalities that they have to offer.
- Configure both the SQL Server Database and the web application so that their "digital footprint" is not revealed.
- Make sure for both the SQL Server Database and web application, never use network port numbers that are normally used. For example, never make use of "HTTPS port 80" or "HTTPS port 443". In these instances, always make use of a higher port number that is not commonly used. Also, be sure to run vulnerability scans on a regular basis to make sure that any unused network ports are not open. If you find that they are open, then they should be closed immediately, as this is a prime way for the cyberattacker to penetrate the SQL Server Database and/or web application.

3

AN OVERVIEW OF THE
SQL INJECTION ATTACK

So far in this book, we have covered a number of key topics as they relate to digital forensics and its impact on cybersecurity. While it has been mentioned that digital forensics is a rather huge field, the main emphasis has been on data that are missing or that have been heisted in a malicious attempt. In cases like these, the main target of penetration is the database, thus attention has been paid particularly to the SQL Server Database. This is the engine that also drives web applications, and any vulnerabilities will also impact the database in tandem. This was also reviewed in detail.

But in this chapter, we take a deeper dive into a particular threat variant that is still used by the cyberattacker that particularly targets the SQL Server Database. This is known as the "SQL Injection Attack".

What a SQL Injection Attack Is?

A SQL Injection Attack can be technically defined as follows:

> SQL injection (SQLi) is a cyberattack that injects malicious SQL code into an application, allowing the attacker to view or modify a database. According to the Open Web Application Security Project, injection attacks, which include SQL injections, were the third most serious web application security risk in 2021. In the applications they tested, there were 274,000 occurrences of injection.
>
> *https://www.crowdstrike.com/cybersecurity-101/sql-injection/*

Put in another way, this is where malicious SQL code can be literally transferred from the front end (which is the web application) back into the SQL Server Database. This is illustrated in Figure 3.1.

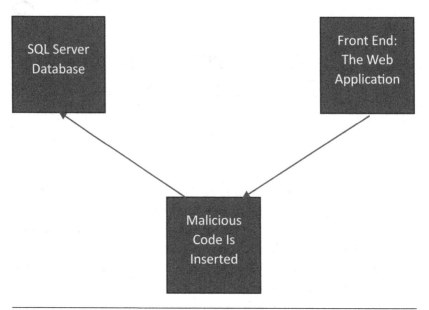

Figure 3.1   The deployment of a malicious payload.

An Example of a SQL Injection Attack

This is an example of how a SQL Server Injection Attack can take place. In this case, suppose that we wish to visit an online store in order to procure products and services. The domain for this situation can be something like this:

https://companyzyz.com/onlinestore

However, suppose we wish to visit the page that has products and services that are listed for a retail price of $200.00 or below. The software development will then create this kind of URL for this very purpose:

https://companyzyz.com/onlinestore/productsservices.
php?val=200

Here is a breakdown of what this URL actually represents:

- https://companyzyz.com: This is the domain of the retailer.
- https://companyzyz.com/onlinestore: This is the URL path to access the overall online store.
- https://companyzyz.com/onlinestore/productsservices.
  php?val=200: This URL path will take you directly to the page or pages that will display all of the products and services that are priced for $200.00 or lower.

- PHP: This is the scripting language that is used to create this specific page. It is an acronym that stands for "hypertext preprocessor".

Now, in order to penetrate into the SQL Server Database, the cyberattacker will merely try to alter the value of "200". To do this, the SQL code is as follows:

https://companyzyz.com/onlinestore/productsservices.php?val=200' OR '1 = '1

With the above URL now entered, it is highly likely that the result will return the pages for all of the products and services at any retail price point. The primary reason for this is that the cyberattacker has actually altered the logic of the SQL query. This is done by making use of the "OR" operator, and having that set to a default value of "1", as this means a syntax of "true" to the SQL Server Database. But in the actual back end, here is how the SQL Server Database actually understands this kind of particular query:

```
SELECT *
FROM ProductsServicesTbl
WHERE Price < '100.00' OR '1'='1'
ORDER BY ProductServiceDescription;
```

Here is a breakdown of the query:

SELECT: This is the command for the SQL Server Database to select a table.

FROM: This part of the query tells the SQL Server Database table to actually look up.

WHERE: Initially, this will tell the SQL Server Database to display all products and services that have a retail price of $200.00 or lower. But the last part of the query will theoretically confuse the database to now return and display all of the products and services listed at any retail price on the online store.

ORDER: This part of the query tells the SQL Server Database to show only those products and services at any particular price point, but the filter being that there must be a textual description that is associated with them.

The above example shows how a malicious SQL code can be inserted directly into the SQL Server Database. But at this point, it is very important to keep in mind that this is just a very simple example in order to demonstrate how a possible SQL Injection Attack can potentially work. The SQL Server Injection Attacks of today are far more complex than this, obviously.

It is also important to point out that this simple example assumes that the web application does not actually confirm the validity of the input provided. In other words, there is no control in place to check for this. Rather, the web application merely assumes that any input is thus valid and safe. But keep in mind that many businesses today have this safeguard in place if their web application is powered by a SQL Server Database as the back end. In other words, launching this kind of threat variant is not as simple as the example demonstrates.

Real-World Examples of SQL Injection Attacks

It should be noted here that SQL Injection Attacks are as old as phishing attacks, which go back all the way to the early 1990s. But given the recent advancements in technology, especially that of generative AI, SQL Injection Attacks have become deadlier, more covert, and, worst yet, much stealthier. Thus, the victim will not even know that they have been impacted until it is far too late to do anything about it. The following are some of the well-known victims of the SQL Server Injection Attack:

1. Guess.com:
   At least 20,000 credit card numbers were hacked into and used to make fraudulent purchases of products and services.
2. Petco.com:
   Through a vulnerability in the SQL Server Database, over 50,000 credit cards were also stolen, and used to make fraudulent purchases of products and services.
3. MasterCard.com:
   To date, this has been deemed to be the largest SQL Server Injection Attack ever launched. In this case, over 40 million credit card numbers were stolen and used to make fraudulent purchases of products and services.

4. Guidance Software:
A cyberattacker altered the SQL Server Database of this particular company, and as a result of this, they were to gain the financial datasets of at least 3,800 clients.

5. TJX:
The SQL Server Databases at TJX were also hacked into, and because of that, the financial transactions of over 1 million datasets were heisted.

6. United Nations:
This has been deemed to be a different kind of SQL Injection Attack, in which datasets were not really heisted. Rather, the threat variant was used to deface the website, by deploying political statements onto it. This kind of threat variant is also referred to as a "Script Kiddie".

The Addition of Programming Languages

Up to the time of the internet bubble of the late 1990s, creating source code, depending upon the kind of application for which it is being developed, was a rather straightforward process. But when this time period occurred, many new software development tools came out that made creating and compiling the source code a lot easier. Some of these have included the likes of Cold Fusion, ASP. NET, Sybase, Informix, WordPress, Joomla, Drupal, PHP, Perl, etc. While this was certainly an advantageous way in order to connect to a SQL Server Database (or for that matter, any kind of database, such as Microsoft Access, Oracle, etc.), there was one main problem with it.

And that is, the software development team did not incorporate a mechanism from within the source code in order to confirm the validity of the inputs that were entered into the web application by the end user. In fact, this trend even continues today, where software developers are now making use of open-source APIs. These are actual libraries of source code but can be modified and/or edited in order to fit the needs and requirements of the web application. These APIs are often used as the bridge that connects the SQL Server Database to the web application, and vice versa.

But the other problem here is that these open-source APIs are often hosted on publicly accessible repositories, and the hosts that sponsor them do not upgrade these open-source APIs as they should. Because of this, whenever they are downloaded by the software development team in order to build out the connections between the SQL Server Database and the web application, these open-source APIs are not tested to make sure that they have been updated. As a result, they possess many flaws and vulnerabilities, that leave the database even more prone to a SQL Injection Attack. This is illustrated in Figure 3.2.

It is also important to note that with these software packages just described, there is no "one size fits all" approach that can be used to create the SQL statements that are needed for the database. Therefore, the software development team has to be well-versed and trained in the particular software package that they will be using to create the source code for the web application. If not, this could lead to even more vulnerabilities, gaps, and weaknesses being created. Also, the source code that is created today for the web application tends to be "dynamic" in nature. In other words, the source code can be automatically corrected and/or adjusted according to the requirements of the web application as they change. Although once again this can prove to be a great advantage, the software development also has to keep in mind that the SQL statements that are incorporated into the database are also equally adjusted and/or corrected as well.

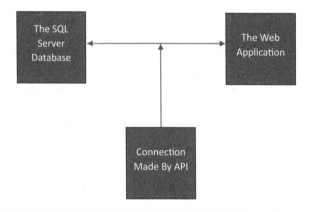

Figure 3.2   An API connection.

How the Backdoors for a SQL Injection Attack Are Created

When the source code and the SQL statements are created for the SQL Server Database, there are a number of areas that can actually offer a backdoor of sorts that cyberattacker can use to gain covert entry and launch a SQL Injection Attack. The following are some examples of this:

1. The dynamic string building:
   This is a specialized technique that allows for the software development team to automatically construct SQL statements during a phase known technically as "runtime". It can be technically defined as follows:

   > Runtime basically means when program interacts with the hardware and operating system of a machine.

   https://stackoverflow.com/questions/3900549/what-is-runtime

In this regard, many software developers choose to construct their SQL statements at "runtime". From here, many different kinds of parameters and permutations can used to create differing SQL statements that can synchronize automatically based on the inputs that are entered by the end user into the web application. For example, since "SELECT" is a key line of source code in a SQL statement, different ones will be created as well, in order to select the right table to be used in the SQL Server Database, which will be contingent upon once again the kind of input the end user provides. But believe it or not, while this technique may be effective, it is far from actually being secure. In order to accomplish this goal, the software development team also needs to incorporate what is known as "parameterized queries". This is where the permutations and boundaries that have been established for a potential set of queries presented to the web application by the end user are actually "embedded" into the SQL statement. As a result of this kind of technique being used, any input that is provided into the web application would not necessarily be interpreted by the SQL Server Database as a command to execute a query that has been submitted by the end user into the web application. Thus, this will greatly mitigate the risk of a SQL Injection Attack in actually happening. But if these "parameterized queries" are not embedded, this will of course greatly increase the risk of a SQL Injection Attack actually occurring.

2. Incorrectly handled escape characters:

In the world of the SQL Server Database, the single quote character, which is represented as a " ' ", is what specifically separates the source code of the database from the actual datasets that it is trying to access based upon the query from the end user. For example, the SQL Server Database will assume that anything after a quote is the actual source code that it needs to execute. Because of this, one can tell pretty quickly if a particular website is prone to becoming a victim of a SQL Injection Attack by merely typing a single quote in the URL. In most cases, if you try this, you will get the following error message:

Warning: SQLServer_fetch_assoc(): supplied argument is not a valid SQLServer result resource

The primary reason why you will get the above error message is that the single quote (" ' ") is viewed as what is known as a "string delimiter". This can be technically defined as follows:

A delimiter is one or more characters that separate text strings. Common delimiters are commas (,), semicolon (;), quotes (", '), braces ({}), pipes (|), or slashes (/\). When a program stores sequential or tabular data, it delimits each item of data with a predefined character.

https://www.computerhope.com/jargon/d/delimite.htm

3. Incorrectly handled query assembly:

Some very complicated applications, such as those that involve data warehousing, need to have the source code written with dynamic-based SQL statements. A good example of this is a multinational company with multiple SQL Server Databases, which stores datasets in literally hundreds of tables. If a part of this web application were to be an actual online store, then you would most likely have many customers accessing their individual accounts in order to check on past purchases or to see what the status is of a pending order for products and services. Obviously, this kind of request will be done over an Internet connection. The primary vulnerability here is that a cyberattacker could conceivably manipulate this request in a malicious way. The end result is that a SQL Injection Attack could transpire in this process, resulting in revealing other private and confidential datasets (such as credit card numbers and bank account information), which could then, in turn, be used to launch deeper and more extensive security breaches.

4. Incorrectly handled errors:

As it has been reviewed earlier in this chapter, whenever an end user of a web application inputs an invalid piece of information, such as an invalid username and/or invalid password, an error message will of course be displayed. This is meant to alert the end user that they have not inputted something correctly. While this is a good thing, but if the software development team has created the source code that will display a detailed message, this can also give the cyberattacker certain clues about the actual mechanics of the SQL Server Database. Therefore, the error messages that are generated should be as generic as possible, but still give enough feedback to the end user as to what they did wrong in this process.

5. Improperly handled submission forms:

When an end user visits a particular web application, the one place where they will most likely enter the most information and data is on the "Contact Us" page. While most of the time one submission of this form will be enough, it could also be the case that it may have to be submitted a number of times. This could happen because of a number of reasons, such as a lost Internet connection or even technical glitches within the servers that host both the SQL Server Database and the web application. Or, it could also be the case that a cyberattacker could be submitting multiple "Contact Us" form submissions during intervals in order to ascertain some of the weaknesses in both the SQL Server Database and the web application.

Because of this, it is a good practice for the software developers to create two kinds of lists:

- A White List: This is the listing that displays what alphanumeric characters can be submitted on the "Contact Us" page in a safe manner.
- A Black List: This is the listing that displays what alphanumeric characters should be submitted on the "Contact Us" page because they could result in a SQL Injection in occurring.

The main problem here is that while this should be the best standard that is practiced, most of the time it is not. As a result, a SQL Injection Attack will most likely happen. The ultimate goal of having these two particular listings in the first place is to make sure that all of the input is properly validated the first time that they are entered in the "Contact Us" page of the web application. Having the "White List" and the "Black List" becomes extremely crucial if the end user has to fill out other kinds of forms that require various types of information and data, such as selecting products and services on an online store where there could be literally hundreds of options to go through and select.

6. Insecure database configurations:

It is very important to keep in mind that your SQL Server Database (or any other kind or type of database for that matter) is configured properly. Although there are many areas to take into consideration when first creating the SQL Server Database, the following can be deemed to be among the most critical:

- Creating the source code:

    At this step, the software developers need to take the time after each module of source code has been created to test it for any gaps or weaknesses. This is best done through either penetration testing or vulnerability scanning, and the importance of this was stressed greatly in the first chapter of this book. Equally if not more important is to test the open-source APIs that have been deployed into the source code, to make sure that they have been updated as deemed to be necessary.

- The access:

  Most SQL Server Database tools come with a set of default accounts that can be used, Typically, these are the "root" or the "administrative" ones. These are primarily reserved for the database administrator. However, extreme consideration has to be done as to how these default accounts should be deployed and allocated. Remember, the SQL Server Database is probably going to be among your most prized possessions in your digital asset portfolio. Therefore, you will want to follow the methodologies of "privileged access management" (PAM) and "least privilege" in this regard. In terms of the former, this deals with those accounts that are super user in nature, as the one that will be assigned to the database administrator. It can be technically defined as follows:

  > Privileged Access Management is a category of cybersecurity solutions that enables security and IT teams to securely manage access for all privileged identities in an enterprise environment.

  > *https://delinea.com/what-is/privileged-access-management-pam*

In terms of the latter, you will only want to give out those rights, privileges, and permissions that are absolutely necessary, and no more than that. In this regard, it is also very important that the software development team create their own accounts for the sole purpose of creating and compiling the source code for the SQL Server Database. The PAM Accounts should never be used, no matter what the circumstance might be.

The Types of SQL Injection Attacks

Now that we have reviewed in some detail the vulnerabilities of the SQL Server Database to a SQL Injection Attack, it is important to review the major types of them that can happen. There are four key ones, and they are as follows:

1. The SQL Code Injection:

   This is what has been primarily reviewed in this chapter. Malicious lines of code are injected into the SQL Server Database when new SQL statements or database commands are inserted. The most common kind of SQL Injection Attack is to attach a SQL Server "EXECUTE" command to the vulnerable SQL statement.

2. The SQL Manipulation:

   This is where the SQL statement is modified through a set of specific operations, which involve primarily altering the "WHERE" and "UNION" clauses in the overall SQL statement. In fact, the most well-known threat variant is to modify the WHERE clause so that it always returns a value of "true" to the end user. Other operations that can be targeted include those of the "UNION", "INTERSECT", and "MINUS".

3. The SQL Function Call Injection:

   This happens when a "Function Call" from another non-SQL Database (such as Oracle or Microsoft Access) is injected into a vulnerable SQL statement.

4. The SQL Buffer Overflow:

   This is actually deemed to be a subset of the "Function Call Injection" Attack. This kind of attack has been reviewed earlier in this book, and the only true way to mitigate this kind of threat vector from actually happening is to have a regular cadence of applying the needed software updates and patches to the SQL Server Database.

How to Mitigate the Risks of a SQL Injection Attack

As it has been stated numerous times throughout this book, no individual or business is 100% immune to a cyberattack. The same also holds true for a SQL Injection Attack, and the only thing in the end that can be done is to simply mitigate that risk from happening in the first place. Here are some key steps to take, across all levels of the SQL Server Database:

1. At the network level:
   - Only allow access to the Internet (especially if you have employees) through a proxy server. This can be technically defined as follows:

     A proxy server is a system or router that provides a gateway between users and the internet. Therefore, it helps prevent cyber attackers from entering a private network. It is a server, referred to as an "intermediary" because it goes between end-users and the web pages they visit online.

https://www.fortinet.com/resources/cyberglossary/proxy-server

- Configure and deploy robust firewall and router rules on the devices themselves.
- Try to harden your network infrastructure as much as possible.

2. At the application level:
   - Make use of well-established cybersecurity frameworks, such as those from the NIST or CISA.
   - As it has been pointed out numerous times throughout this book, maintain a regular schedule of deploying the needed software patches and updates.
   - Disable any type of stored SQL Procedure Call. An example of this is "xp_cmdshell".
   - Avoid making use of long URLs as you create and develop the web application. A good rule of thumb is to keep at no more than 2,048 characters.
   - Make sure that only valid information and data are inputted by the end user.
   - As stated earlier in this chapter, use only short and generic types of SQL Server Error Messages.
   - Maintain a strong and robust password security policy. Create passwords that are long and complex, and in this regard, make use of a good password manager.
   - Make sure to keep detailed documentation on the following:
     - The SQL Server Database Accounts that have been created, disabled, or even deleted.
     - Stored procedures.
     - Any kinds or types that the use cases that you have created a SQL Server Database for.
   - Always make sure to conduct regular risk assessments and audits of not only your SQL Server Database but also anything that is networked or even associated with it.

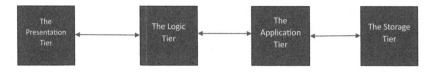

Figure 3.3   The different network tiers.

Finally, it is also highly recommended that you try to create and deploy a four-tier approach when creating both the SQL Server Database and the web application. This is illustrated in Figure 3.3.

These components are described as follows:

1. The presentation tier:
   This is where the "graphical user interface" (GUI) resides for the web application. This is what the end user interfaces with and is considered to be the front end of the web application.
2. The logic tier:
   This is the component in which the actual source code resides that was used to create the web application. For example, this can include the likes of PHP, C#, Java, etc.
3. The application tier:
   This is where all of the open-source APIs reside. Once again, it is very important for the software development team to confirm that they have been upgraded before they are released into the Production Environment.
4. The storage tier:
   This is where the actual SQL Server Database resides and contains all of the needed datasets.

By taking this kind of multi-tiered approach, the chances of having a SQL Injection Attack are also greatly mitigated. In other words, the statistical probability of the cyberattacker actually penetrating into the SQL Server Database becomes significantly lower.

4

CONCLUSIONS

Using Generative AI in Digital Forensics

So far in this book, we have covered numerous topics, and to summarize, they are detailed as follows:

- Major security breaches, most notably those on the critical infrastructure on a global basis.
- The importance of both penetration testing and threat hunting, and its strong relationship with digital forensics.
- An overview of digital forensics.
- How digital forensics can be used to investigate security breaches that have involved data deletion or data exfiltration.
- A detailed overview of the SQL Server Database.
- A technical review into SQL Server Injection Attacks, which is the prime threat variant of today that is impacting the SQL Server Database.

In this last chapter, we conclude how generative artificial intelligence (AI) can be used in conjunction with digital forensics. It is very important to keep in mind that the evolution of generative AI is still ongoing, and therefore, its incorporation into the field of digital forensics and the tools and software packages that are used is still new. But before we do an overview of this, it is important to provide an overview of the field of "AI".

What Is Artificial Intelligence?

We have written and published three books on this topic, and the titles for them are as follows:

- Practical AI for cybersecurity
- Generative AI: phishing and cybersecurity metrics
- Generative AI and cyberbullying

DOI: 10.1201/9781003469438-4

You can refer to these books for both an extensive and exhausting review of generative AI. But it is important to note that generative AI is actually a subfield of AI in general. The primary goal of AI is to mimic the thought-making and reasoning powers of the human brain and use that to make future predictions for just about every industry and application that is imaginable. But, keep in mind that the human race is still very far away from understanding the human brain in its entirety, and it is quite possible that we never will fully understand it.

But for purposes of this chapter, there are three key areas of AI that have started to and will continue to play a very pivotal role in the world of digital forensics. These are described in the next four subsections.

Machine Learning

From a technical perspective, machine learning (ML) can be defined as follows:

> Machine Learning, often abbreviated as ML, is a subset of artificial intelligence (AI) that focuses on the development of computer algorithms that improve automatically through experience and by the use of data.
>
> *https://www.datacamp.com/blog/what-is-machine-learning*

ML has been around for the longest time, going back as far as the 1960s. But over time, many advancements have been made to it, and it is the kind of AI model that can be viewed literally as "garbage in and garbage out". A simple example of this can be seen in Figure 4.1.

As you can see from the illustration, the datasets are first inputted or "ingested" into the ML model. From there, they are processed by the associated algorithms that have been programmed, and the

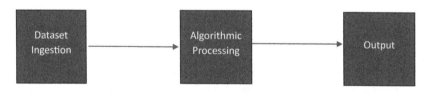

Figure 4.1   The processing in AI.

output is generated. This is in fact, deemed to be the simplest form of an AI model. However, it is very important to note here that the robustness of the output is only as good as the datasets that have been fed into it. Therefore, special care must be taken that they are cleansed and optimized to the fullest degree that is possible.

Neural Networks

The next step up from the ML model is known as the "neural network". This too can be technically defined as follows:

> A neural network is a method in artificial intelligence that teaches computers to process data in a way that is inspired by the human brain. It is a type of machine learning process, called deep learning, that uses interconnected nodes or neurons in a layered structure that resembles the human brain.
>
> *https://aws.amazon.com/what-is/neural-network/*

In this case, and as opposed to the ML model, there are multiple layers that are involved in a neural network model. The primary component of this is what is known as the "neuron". This is the fundamental cellular structure that powers the entire human brain. There are billions of them, and in fact, it is estimated that there are as many as 100 billion-plus neurons in the human brain. Figure 4.2 illustrates a neural network model.

As you can see from Figure 4.2, multiple layers are used in order to fully process the datasets in order to yield the output. In order to replicate the neuron, different statistical weights are thus assigned.

Computer Vision

The next major subfield of AI is known as "computer vision". It can be technically defined as follows:

> Computer vision is a field of artificial intelligence (AI) that uses machine learning and neural networks to teach computers and systems to derive meaningful information from digital images, videos and other visual inputs.
>
> *https://www.ibm.com/topics/computer-vision*

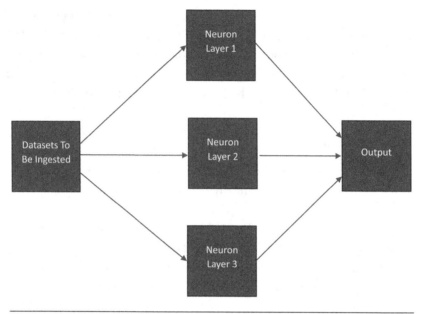

Figure 4.2   A Neural Network.

This is still an emerging field in AI, and as you can see from the above definition, the goal is to replicate the vision process of the human eye. More specifically, the goal is to try and replicate how the human eye sends information to the optic nerve, which is then connected to the optic disc. This, in turn, is connected to the human brain, for processing visual information and data.

As of now, one of the biggest applications of computer vision is CCTV technology. These are cameras that are deployed in whatever application is needed for them. A biometric modality, known as "facial recognition", is used for further analysis of the subject in question, in order to 100% confirm their identity. In the future, computer vision can play a key role in digital forensics, especially when it comes to analyzing and capturing videos and images of the evidence on a real-time basis, which could in the end save an enormous amount of time for the digital forensics investigators.

Generative AI

This is the hottest and most recent trend in the world of AI. This has been primarily due to the birth of "ChatGPT", which was developed by OpenAI, with a large amount of help from Microsoft. It can be technically defined as follows:

> Generative AI is a type of artificial intelligence technology that can produce various types of content, including text, imagery, audio and synthetic data.

> *https://www.techtarget.com/searchenterpriseai/definition/generative-AI*

This is illustrated in Figure 4.3.

There are some key differences between generative AI when compared to the other AI models. They are as follows:

- The GPT4 algorithm is primarily used. It is an acronym that stands for "generative pre-training transformer".

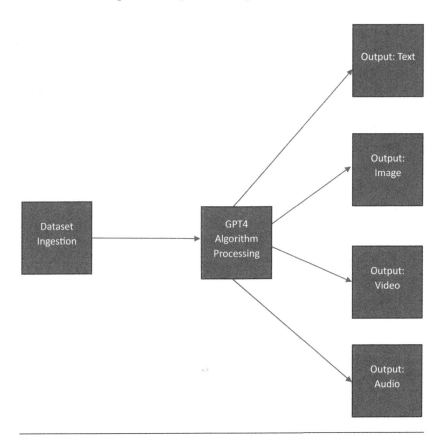

Figure 4.3 The GPT algorithms.

- Different kinds of outputs can be created, depending on the query that is submitted to the generative AI model.
- The generative AI model can also create original content on its own.
- It can also produce what is known as "synthetic data", which is essentially fake data that looks like a real dataset so that a generative AI model can initially train on it if no other original dataset can be found. This is also used to create "deepfakes", which are artificially created videos and images of real-life people, and is used primarily for nefarious-based purposes, such as launching a phishing attack or even a social engineering attack.

The Role of Generative AI in Digital Forensics

Here are some of the key ways in which this can happen:

1. Automated log analysis:
   This will play a huge role in cybersecurity. For example, depending on the total number of network security tools that are used, a lot of log files can be generated. It is an almost impossible task for an IT security team to go through all of them, in order to parse out the pertinent information and data. Therefore, generative AI can be used to automate this process and find key pieces of evidence in just a matter of minutes, such as unusual and anomalous spikes in network traffic behavior.
2. The detection of malware:
   This kind of malicious payload is used in primarily launching ransomware attacks. But here, generative AI can be used to examine key pieces of digital evidence in an effort to construct what future malware could potentially look like.
3. Image and video analysis:
   Once again, this is where computer vision will play a key role. For example, generative AI algorithms can be used to comb through large amounts of multimedia content, including the identification of faces, objects, or text within images and videos, which are crucial pieces of evidence.
4. Natural language processing:
   This is more commonly referred to as "NLP". When used in conjunction with digital forensics, it can rapidly process and analyze any kind or type of text or written data.

This can then be used to identify and ascertain recurring phrases, abnormal patterns, and covert connections between individuals.

5. The network traffic analysis:
   Apart from merely analyzing log files, generative AI can be used to collect evidence on a real-time basis as it monitors network traffic between all of the endpoints in question. It can also be used to identify deviations from usual network traffic patterns and alert the IT security team when an event needs to be further investigated. Generative AI can also be used to correlate suspicious network events with known attack patterns, providing valuable pieces of evidence for the digital forensics team.

6. The triaging:
   In a full-force digital forensics investigation, there will, of course, be a lot of information and data that will be collected. In this regard, generative AI can be used to analyze the metadata, other types and kinds of content, and various data-related attributes. Further, this will allow the digital forensics team to quickly identify and focus on the most important evidence earlier, thus allowing for the effective triaging of the most critical pieces of evidence.

7. Prompt engineering:
   Simply put, this is the art of crafting a query for the generative AI algorithm so that it can create the most robust output that is possible. It can also be applied to digital forensics as well. For example, well-created prompts can the digital forensics investigators to very quickly sift through vast amounts of data, using standard English-based keywords.

8. Filling in the gaps:
   In a complete digital forensics investigation, there will be a lot of extrapolation that needs to be done. But many times, it could be difficult to "connect the dots", if any pieces of related evidence are actually missing. Therefore, generative AI can thus be used to create the "synthetic data" that can be used to create fake, but real-looking pieces of evidence in order to paint a complete picture to the digital forensics team. This will prove to be especially crucial when they present their case to a court of law.

9. Image reconstruction:

In this regard, generative AI can be a great boon in order to reconstruct what is known as "latent evidence". This is technically defined as follows:

> The word latent implies that the prints are hidden or not easily seen without help (either chemical, physical, photographic, or electronic development).

Thus, generative AI can be used to take out the guesswork if there are any fingerprints, or even images that are left behind and are badly damaged. Assuming that the model has been fed the right kinds of datasets, the generative AI model should be able to reconstruct and compensate for the damage in the fingerprint and/or image. This will prove to be the most accurate way of doing this as opposed to the traditional methods described in the above definition.

Index

Printed in the United States
by Baker & Taylor Publisher Services